STRUCTURE,
CONTEXT,
COMPLEXITY,
ORGANIZATION

STRUCTURE, CONTEXT, COMPLEXITY, ORGANIZATION

Physical Aspects of Information and Value

Karl-Erik Eriksson
Kristian Lindgren
Bengt Å. Månsson

World Scientific
Singapore • New Jersey • Hong Kong

Published by

World Scientific Publishing Co. Pte. Ltd.
P.O. Box 128, Farrer Road, Singapore 9128

U. S. A. office: World Scientific Publishing Co., Inc.
687 Hartwell Street, Teaneck NJ 07666, USA

Library of Congress Cataloging-in-Publication data is available.

STRUCTURE, CONTEXT, COMPLEXITY, ORGANIZATION

ISBN 9971-50-023-4

Printed in Singapore by Koo Printing Pte Ltd.

FOREWORD

A physicist getting involved in broad interdisciplinary collaboration soon finds that the systems studied in other fields of knowledge, such as biology or economics, are in many respects more complicated and more fuzzy than the systems usually studied in physics. Concepts and ideas developed there, which at first may appear strange to him, can after some time turn out to be quite adequate for the systems studied. Still, he may experience that very fundamental physical concepts are relevant to a wide variety of systems and processes—not only in principle, but also in practice. The basic concepts of thermodynamics and of information theory are general enough to play such a role.

The authors have been involved in studies of the handling of natural resources in human societies. There we met problems and ideas which led us to the theme of this book: a perspective and a set of concepts, useful for describing and understanding processes in which structure emerges.

This book had not been possible to write without the broad discussion, involving many disciplines, going on within the University of Göteborg and Chalmers University of Technology which together form a free and open intellectual environment. We cannot here mention single persons or research groups.

More locally we have an intellectual home in the Physical Resource Theory Group (belonging to both universities), and we thank Gun Fornell (secretary), Saiful Islam (now in München), Tomas Kåberger, Sten Karlsson, Giorgio Peressutti, and Göran Wall for a fruitful daily collaboration.

Karl-Erik Eriksson also thanks Bo-Sture Skagerstam for collaboration, the result of which is reviewed in this book. Ulf Norrsell is thanked for an illuminating discussion on modern aspects of the old mind-body problem.

He also thanks Tito Arecchi, Hermann Haken, and Stig Lundqvist for fruitful collaboration in the planning of the Adriatico Conference on the Physics of Structure and Complexity in Trieste 1986 and Abdus Salam for giving us the opportunity to hold this conference at the International

Centre for Theoretical Physics.

Important parts of Bengt Månsson's work for this book were done when he was visiting the Department of Chemistry of the University of Copenhagen and the corresponding department of the University of Chicago. For hospitality and for enlightening discussions he is grateful to Thor Bak and Preben G. Sørensen as well as to R. Stephen Berry.

For this work we have received financial support from The Swedish Council for Planning and Coordination of Research, The Swedish Natural Science Research Council, and the program of Nordic Research Courses (Nordic Council of Ministers).

Göteborg, October 1987

Karl-Erik Eriksson Kristian Lindgren Bengt Å. Månsson

CONTENTS

INTRODUCTION

Our universe, as a closed system, obeys the second law of thermodynamics, i.e., its entropy, its disorder, increases. Still, in the midst of this we are witnessing an opposite process, in which order is emerging. The picture we have is that the universe has evolved from a totally unstructured state of matter and radiation to a highly structured state. On the surface of our Earth, some of this structure takes highly sophisticated forms in living organisms and their complex interaction and in the social as well as material organization of human societies. What makes all this possible, thermodynamically, is, of course, the fact that over the universe as a whole the entropy increase is large enough to compensate for local creation of order.

The aim of this book is to sketch a perspective and to present a set of concepts, which can be useful in order to describe and understand processes in which structure emerges. The word 'describe' gives a hint that 'information' is an important concept. Information is a general concept, which can be applied not only to the description of a system but also to the described system as well. We shall use the concept as it was introduced by Shannon in information theory [Shannon 1948, Shannon and Weaver 1949].

Then one may ask for the origin of structure (measured in terms of information) in the universe; one may ask how it is being transformed and how creative and destructive processes—structure formation and entropy production—are related to each other.

Many ideas and concepts have been developed which make information a very flexible and versatile theoretical tool. Several of those ideas were present already in Shannon's original work.

One very important idea, which ties all of information theory firmly to physics, is Jaynes' idea [Jaynes 1957, 1963] that statistical mechanics is simply an application, albeit a very fundamental one, of information theory. Conversely, this idea has made the method of statistical mechanics available to information theory itself. This has been very fruitful and has opened several fields of application, not only in the natural sciences and engineering [Levine and Tribus 1979], but also in the social sciences [Wilson et al. 1981, Weidlich and Haag 1983, Weidlich 1987].

Information theory is directly applicable to the analysis of language. A text may be viewed as a finite one-dimensional lattice with one letter (in a generalized sense, including punctuation marks and space) per lattice site. The information content in such an expression may be analyzed in several ways, for instance with respect to the information content of correlations among pairs, triplets etc. of letters (correlational information) and information modulo those correlations (textual information) [Shannon 1948, Brillouin 1956]. This and other kinds of analysis may be generalized to several dimensions and to the case of a continuum.

Textual information describes individual or local features and is clearly necessary in order to recognize something as different from something else. In order to identify a *context*, it is necessary that such individual and local features have some degree of permanence. A context is thus characterized by permanent or semi-permanent textual information. Complexity of a system, on the other hand, is related to its internal structure, i.e., to correlations; we shall discuss this below.

A useful concept, most readily applicable in the continuous case, is that of *resolution*, usually given as a length (uncertainty in position) [Chaitin 1979, Eriksson and Lindgren 1987]. If an object in *n* dimensions is *d*-dimensional, then *nd* coordinates are needed to describe its position. If the resolution is improved by a factor of two, then the description is extended by one more bit per coordinate; the information is increased by *nd* bits. In this way a generalized concept of *dimension* [Mandelbrot 1983] has been defined related to information. This dimension concept is particularly useful in the case of self-similar structures.

One can use resolution to travel in length scale. At infinitely bad resolution one sees only the gross properties of a system. As the resolution improves one is able to see first large-scale structures and then structures on an increasingly finer scale, until one reaches perfect resolution. In this way it is possible to localize information in scale.

A concept in applied thermodynamics which has proven to be quite fundamental is *exergy*, the amount of mechanical work which could be (reversibly) extracted from a system, e.g., [Brønsted 1955, Rant 1956, Tribus et al. 1966, Procaccia and Levine 1976, Smith-Sørensen 1977,

Levine and Tribus 1979]. An expression for this quantity was given already by Gibbs [1873]. Its main advantage is that it is so constructed that it takes into account both the first and the second laws of thermodynamics, i.e., that energy is conserved and entropy is produced. The exergy is thus zero at thermodynamical equilibrium but positive, whenever there is a deviation from equilibrium. The exergy concept has a long history in engineering literature, but it is only recently that it it has come to be widely accepted and applied internationally. Exergy is closely related to *Kullback information* [Kullback 1959] (also called relative information). The reference state to be used is then the equilibrium state [Jaynes 1963, Tribus et al. 1966, Lindblad 1983].

According to the standard model of cosmology our universe was at least once in thermal equilibrium. The idea of non-equilibrium and hence also exergy as arising from the expansion of the universe, was present already in Tolman's classical work on cosmology and thermodynamics. The decrease in pressure and temperature due to the expansion changed the conditions for equilibrium, and the equilibration processes could not keep pace with the expansion. More modern discussions of this question were given by Davies [1974], Landsberg [1978], and Frautschi [1982]. A specific model for the creation of the nuclear exergy was given by Eriksson et al. [1982].

As a result of the expansion and of the matter distribution of the universe becoming thinner, at some stage the pressure became too low to prevent gravitational attraction from clumping matter together.At this stage the state of evenly distributed matter became unstable. The universe had to form the very uneven distribution of a whole hierarchy of structures: clusters of galaxies, galaxies, galactic centres, clusters of stars, stars, and planets.

The cosmic evolution indicates a general feature of self-organization processes. The value of some crucial physical quantity is changing. The initially homogeneous state stays stable until this quantity reaches a critical value [Turing 1956, Glansdorff and Prigogine 1971, Haken 1977, Nicolis and Prigogine 1977]. At this value there is a bifurcation, at which several inhomogeneous states come up as more stable than the homogeneous one.

The system chooses one of the inhomogeneous states, and the choice is triggered by noise, i.e., it depends on uncontrollable minute details of internal fluctuations or external disturbances [Glansdorff and Prigogine 1971]. Often a whole series of critical values occur, each leading to a new uncontrollable choice.

One way to describe this is that information at the microscopic level (noise) determines the choice, when part of the evenly distributed exergy is converted into the information contained in the structures that are formed in the process. Entropy production can be viewed as the disappearance of information into the microscopic end of the scale, where it is irrevocably lost into random fluctuations. Thus, in a self-organizing system information originally present on a scale of the size of the system travels towards the microscopic scale, where it disappears. It halts along the way to form permanent or semipermanent structures on intermediate scales. The exact forms of such structures depend on random information coming from the microscopic end of the scale.

This description is generally true for spatially structured self-organizing systems. In particular, if the standard model of cosmology [Weinberg 1972] is correct, it is true of the universe itself. Some time ago, in a popular article on the cosmic energy conversion, Dyson [1971] described how several 'hang-ups' make the entropy production of the universe proceed relatively slowly. One hang-up is the size of the universe which makes motion over cosmic distances take a long time. Another one is in the nuclear burning in the stars, where weak interaction is a necessary intermediary step. Thus a picture emerges of a universe with an inflow of exergy at the cosmic end of the scale, due to the expansion of the universe, and a loss of exergy at the microscopic end of the scale, due to entropy production. On intermediate scales structures are formed, some of which are quite long-lived. At some place in such a scheme one finds the Sun and the Earth and the exergy flows and conversions at the Earth's surface, including the material/energy cycles of the Earth, the metabolisms of living beings, and the conversions of energy and materials in human societies [Boulding 1978].

In all this there are complex networks of flows of energy and materials; there are even networks within networks. Systems of this kind may be subject to information-theoretical analysis. Basic ideas in this context have been formulated in ecological theory [Ulanowicz 1981]. The same method can also be applied to input/output descriptions of economic systems [Leontief 1951, Kneese et al. 1970]. In this way one can define organizational measures to account for evolutionary level/specialization and robustness/redundancy, admittedly crude measures, but relevant, at least as a step in the development of concepts.

Information is recorded in systems on the Earth's surface, such as in geophysical structures and in genetic codes. Part of the information carried by an evolving systems is information on how the system can be maintained/reproduced. A dried-up river is prepared to guide water from coming rains. A hypercycle of organic macromolecules, materialized in concentrations above certain threshold values, can determine the reaction pathways for inflowing organic material. The instructive character of genetic information is even more obvious. In society, ideas, practices, and various traditions, such as cultural or scientific traditions, play similar roles.

In this way one part of the system functions as a description or blueprint for the whole system; the system carries reference to itself. Bateson [1972] has shown that such self-reference may play a crucial role in learning processes and probably in consciousness itself. Self-reference implies a transcendence of the boundary between one logical level and its metalevel [Russell and Whitehead 1910-13]. This has been vividly expressed in M. C. Escher's artistic works. His pictures were interpreted by Hofstadter in his great self-referential essay on self-reference "Gödel, Escher, Bach" [Hofstadter 1980]. Self-reference through numbering of propositions was the method used by Gödel [1931] to construct his proof that any axiomatic calculus, sufficiently large to include number theory, must contain statements which are true but impossible to prove within the same calculus.

A nice formal way to consider the production of one system (the product) by another (the machine) according to a blueprint, is to think of the machine as a computer with the blueprint as input and the product as output. In the case of a one-dimensional lattice this is a good picture even in the literal sense. Turing introduced the concept of the universal computer [Turing 1936], a simple Gedanken-device with the capacity to carry out any finite computation. Any system with a finite set of available states may be represented by a binary string A. Sometimes such a string may be represented by a simplified description, for instance when it consists of only zeroes or repeated strings or when it is of the form A=0101101110111101... . The simplified description may then be a program for a Turing machine which then produces A as an output. Such a program P can be viewed as an algorithm (blueprint or recipe) for constructing the system in the wanted state A. The length of the shortest program P_0 which can achieve this is called the *algorithmic information* [Kolmogorov 1965, Chaitin 1966] of A. The number of steps needed for the machine to produce A from P_0 is the *depth* [Bennett 1986] of A. These concepts are widely applicable since they only require that the system can be described by (or coded into) a sequence of zeroes and ones. However, a consequence of Gödel's theorem is that there does not exist a general algorithm (or procedure) for finding the shortest algorithm for A. For most long strings A containing correlations, the algorithmic information is well approximated by the textual information—a computable quantity. We shall use this fact to build upon an idea due to Bennet [1982] and express the entropy of a physical system in the thermodynamical limit as the mean value of textual information.

A computer has to be a material physical system, functioning according to physical laws [Landauer 1987]. The discussion of the minimal costs for computation in terms of exergy loss is of great principal importance. A structure generating process can be viewed as a non-deterministic computation process executed by a 'thermodynamical' Turing machine. By combining these aspects we could get an understanding of the minimum thermodynamical costs to operate structure-generating systems.

A particularly useful theoretical laboratory for studying discrete dynamics is the cellular automaton [Wolfram 1983]. A cellular automaton is a lattice of one or more dimensions with a discrete number of states available at each lattice site. It evolves in time by discrete steps according to fixed deterministic or stochastic rules. Cellular automata can be used to model physical systems. The Ising model is the most well-known example of this.

To work with new concepts and ideas can also give a fruitful impact on an old field. Thus, one topic which we deal with in this book is the open quantum-mechanical system. The bifurcation process known as quantum measurement can be understood in terms of the statistical properties of an open quantum-mechanical system [Eriksson 1987].

Values as discussed and applied in daily life, in philosophy, or in economic theory are not totally immaterial. Moral and economic values may be closely tied to very substantial flows of natural resources and to the way the resource handling systems are embedded in their biological and geophysical environments. The whole question of the relationship between value and its physical basis [Georgescu-Roegen 1971] is quite complex.

If we start at the simplest end, thermodynamics, it seems clear that exergy is a measure of value. There is a far-reaching equivalence, or convertibility, between energy and material resources in this context. In, e.g., processing industries exergy from some fuel or from electricity is used to improve the quality of the processed material. The simplest aspect of the processing of some raw material may be the chemical one, the change in composition. The next aspect may be a structural one, for instance, how a solid material is composed of fibers or crystals of various size [Ayres 1978], and within what limits parameters describing this are allowed to vary. Next is the shape of components and of ready-made products. All this can be suitably dealt with using information theory, in such a way that the relations to thermodynamics are clear.

At room temperature, $kT \ln 2$ is approximately 3×10^{-21} J. This is the 'natural' unit of exergy connected to one bit of information. For most processes the exergy used to transfer one bit of information is much larger. Berry [1972] estimated the exergy content in the shape of a car door and found it to be extremely small. The dissipation in the actual production process is clearly many orders of magnitude larger. Such discrepancies between ideal and (presently) realistic processes does not prove that the theory is uninteresting. Rather, they call for a careful analysis of what determines the order of magnitude of the non-idealities. Our willingness to accept such discrepances shows that we attribute a much higher economic value to structural information, reproducing a prescribed structure, than we do to exergy itself.

In the end what economic value is, or should be, is based on some physically describable service [Ayres 1978] given to one or several human beings. Such a service should lie within certain physical limits, chosen in order not to conflict other values, for instance values connected to other living beings or the freedom of choice of future generations.

To life, in particular to conscious life, is ascribed a fundamental val A special value is attributed to the self-conscious human life. There do n yet exist any good information-theoretical characteristics of life, consciousness, or self-consciousness. Some concepts such as Chaitin's complexity measures [Chaitin 1979] or the ideas of self-reference have clearly a potential to be useful in this context.

To value *only* human life and consider less sophisticated living forms as devoid of value, would be inconsistent, considering our relationship with them. After all, we share with them not only the Earth but also ancestors who were more humble than we are but who had the potentiality to give existence to us. Going back beyond this, the heat engine Sun-Earth-Space carried life as a potentiality. So did long before that the expanding primordial matter/radiation soup, which once constituted our universe.

In this book we deal with concepts and methods belonging to the lower levels in a hierarchy of information theory concepts. These concepts take into account some necessary aspects of life, including human life, but they are far from sufficient to capture the full concept of life.

The ideas used in this book have been collected from many fields, but we have no ambition to review the fields from where we have picked our material, and our collection of references is not systematic. We have mixed old material with new material, elementary discussions with more advanced, in order to make the book as a whole more self-contained. Since the book has not been written for a well-defined category of readers, we hope that in this way it will be more useful. Our aim has been to sketch a perspective on the great process of structure formation going on all the time around us, and to convey some of the most basic theoretical ideas which are useful for work in this perspective.

1. INFORMATION THEORY

Information theory, dealing with information as a quantitative concept, started in 1948 [Shannon 1948]. From the beginning it was close to physical theory [Brillouin 1956], with entropy as a common concept. A few years later Jaynes gave his formulation of statistical mechanics in information-theoretical terms [Jaynes 1957]. This work made the methods and concepts of statistical mechanics available to statistical theory in general. They found applications in many areas [Levine and Tribus 1979], also in quantitative social science [Wilson et al. 1981, Weidlich and Haag 1983, Weidlich 1987]. Information theory itself was quite soon saturated with a few important basic ideas and concepts. When inquiries into the nature of complex systems became a new frontline in physical research—stimulated by highly improved capacity for making computer simulations—then information theory was ready also to provide more sophisticated concepts, e.g., [Grassberger and Procaccia 1984, Procaccia 1985, Grassberger 1986a].

The aim of this chapter is to give a formulation of the fundamenta of information theory which is suitable for statistical mechanics and physical theory in general.

1.1 Basic Concepts

The unit of information, one binary unit or bit, is is the information carried by one binary figure with unbiased a priori probability for being 0 or 1. A number consisting of N binary figures (with unbiased and uncorrelated a priori probabilities) carries N bits of information.

Let us assume that an unbiased n-figure (base = n = integer) carries I_n bits of information. Then a number consisting of q n-figures (unbiased and uncorrelated) carries qI_n bits of information. There exists a unique integer r such that $2^{r-1} < n^q \leq 2^r$, and hence

$$\frac{r-1}{q} < {}^2\log n \leq \frac{r}{q} \,. \tag{1.1}$$

Our number can thus be written with r binary figures but not with $r-1$. Hence

$$r-1 \text{ bits} < q\, I_n \le r \text{ bits}. \tag{1.2}$$

By choosing larger and larger q we find more and more narrow intervals in (1.1) and (1.2). We thus conclude that

$$I_n = {}^2\log n \text{ bits} = \frac{\ln n}{\ln 2} \text{ bits}. \tag{1.3}$$

A number consisting of q n-figures (unbiased and uncorrelated) is an event among n^q possible events, each with the same a priori probability

$$p = n^{-q}. \tag{1.4}$$

The information content I is

$$I = q\, I_n = {}^2\log \frac{1}{p} \text{ bits}. \tag{1.5}$$

The information content in any event which has an a priori probability p, $(0 < p \le 1)$, is defined as

$$I = {}^2\log \frac{1}{p} \text{ bits} = \ln \frac{1}{p}, \tag{1.6}$$

where we have introduced the convention that 1/(ln 2) bits (often called one 'nit,' natural instead of binary logarithm) is equal to unity (thus 1 bit=ln 2= = 0.693).

Consider a system which has n states (microstates) available. We label the states by an index j ($j = 1,\ldots, n$). When complete information is not available on the system one can try to assign a probability p_j to each of the j

microstates,

$$\begin{cases} p_j \geq 0, \quad (j = 1, \ldots, n); & (1.7) \\ \sum_{j=1}^{n} p_j = 1. & (1.8) \end{cases}$$

The probability distribution

$$P = \{p_j\}_1^n = \{p_1, \ldots, p_n\} \tag{1.9}$$

satisfying (1.7) and (1.8) then defines a macrostate of the system.

1.1.1 *Entropy and negentropy*

Unless $p_j = 1$ for some j, the information (about the system) contained in the macrostate P is incomplete. If we should somehow learn that the system is indeed in the microstate j, then our information gain according to (1.6) would be $\ln(1/p_j)$. Thus if the system were in state j our possible information gain would be $\ln(1/p_j)$.

We do not know which microstate is realized but we may still use the p_j's as weights to determine the expected information gain. This is

$$S[P] = \sum_{j=1}^{n} p_j \ln \frac{1}{p_j} \,. \tag{1.10}$$

S is known as the entropy connected to the macrostate (1.9). It is a measure of the lack of information. Clearly S is nonnegative; S vanishes when $p_j = j$ for some j, and S reaches its largest value, $\ln n$, for $p_1 = p_2 = \ldots = p_n = 1/n$,

$$0 \leq S \leq \ln n \,. \tag{1.11}$$

If there are restrictions on the probability distribution (1.9),

$$F_k(P) = 0 \quad (k = 1,\ldots, r), \tag{1.12}$$

then the maximum value of entropy $S_{\max}$ may be smaller than $\ln n$. In this case

$$0 \leq S \leq S_{\max} \,. \tag{1.13}$$

The problem of maximizing (1.10) under the restrictions (1.7), (1.8), and (1.12) will be dealt with later in this chapter.

It is convenient to define also *negentropy* Σ [Schrödinger 1944] such that

$$S + \Sigma = S_{\max} \,. \tag{1.14}$$

Then Σ is confined to the same interval as S,

$$0 \leq \Sigma \leq S_{\max} \,. \tag{1.15}$$

Clearly, Σ vanishes, when the lack of information is maximal, and it has its maximum for zero lack of information. Thus Σ is a measure of the information content of the macrostate P. Lack of information (presence of information) is often connected to disorder (order). Therefore entropy and negentropy are measures of disorder and order respectively.

For two uncorrelated systems, described by macrostates P and Q, entropy—and hence also negentropy—is additive. The combined system has microstates labelled by index pairs (j, k) with probability assignments $p_j q_k$. The entropy of the combined system is

$$\begin{aligned} S[PQ] &= \sum_{j=1}^{n} \sum_{k=1}^{m} p_j \, q_k \ln \frac{1}{p_j \, q_k} = \\ &= \sum_{j=1}^{n} p_j \ln \frac{1}{p_j} + \sum_{k=1}^{m} q_k \ln \frac{1}{q_k} = S[P] + S[Q], \end{aligned} \tag{1.16}$$

and thus additivity has been demonstrated.

1.1.2 *Contrast* (*Kullback information*)

Consider a system A with n possible microstates $j = 1,\ldots, n$. Suppose that, on whatever information we had available, we have made a probability assignment $P^{(o)} = \{p_j^{(o)}\}$, defining a particular macrostate. Assume now that we learn that this probability assignment was prejudiced and that $P = \{p_j\}$ is a better probability assignment (macrostate) for A. The expectation value of the information increase (calculated on the grounds of the better probability assignment) is then

$$K[P^{(o)}; P] = \sum_{j=1}^{n} p_j \left(\ln \frac{1}{p_j^{(o)}} - \ln \frac{1}{p_j} \right) = \sum_{j=1}^{n} p_j \ln \frac{p_j}{p_j^{(o)}} \,. \tag{1.17}$$

This quantity has been called 'relative information', 'Kullback information' [Kullback 1959], or 'surprisal'. We shall call it *contrast*. An important property of the contrast is that it is positive semidefinite,

$$K[P^{(o)}; P] \geq 0, \tag{1.18}$$

with equality only for coinciding macrostates,

$$K[P^{(o)}; P] = 0 \Leftrightarrow p_j^{(o)} = p_j\,, \quad (j = 1,\ldots, n). \tag{1.19}$$

To show this we use the simple function, defined for $x > 0$,

$$g(x) = x - 1 - \ln x\,, \tag{1.20}$$

which has the properties

$$\begin{cases} g(x) \geq 0; \quad g(1) = 0; \\ g'(1) = 0; \quad g''(x) > 0; \quad g''(1) = 1; \\ \lim\limits_{x \to +0} g(x) = \infty; \quad \lim\limits_{x \to \infty} g'(x) = 1. \end{cases} \tag{1.21}$$

As we shall see, this function plays an important role in statistical mechanics and thermodynamics. The graph is shown in Fig. 1.1.

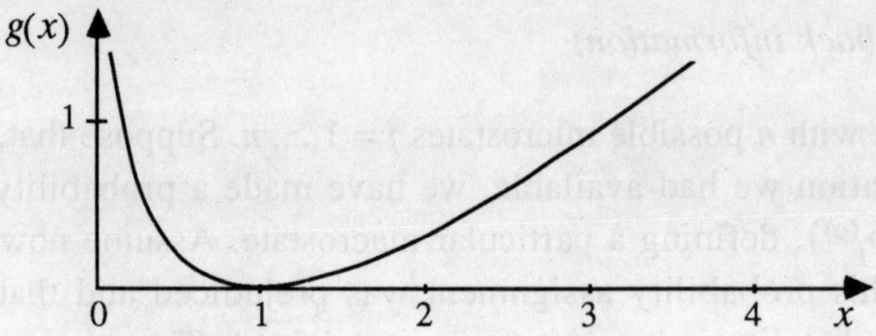

Fig. 1.1. The function $g(x) = x - 1 - \ln x$.

Since $P^{(o)}$ and P satisfy (1.7) and (1.8), we see that (1.17) may be rewritten

$$K[P^{(o)}; P] = \sum_{j=1}^{n} p_j \, g(p_j^{(o)}/p_j). \tag{1.22}$$

Then the properties (1.18) and (1.19) follow immediately from (1.22) and (1.21).

If P is close to $P^{(o)}$, then we can introduce auxiliary variables z_j, such that

$$p_j = p_j^{(o)}(1+z_j); \quad |z_j| << 1; \quad \sum_{j=1}^{n} p_j^{(o)} z_j = 0. \tag{1.23}$$

If we express $K[P^{(o)}; \{p_j^{(o)}(1+z_j)\}_j]$ as a power series in z_j, then to second order in the z's ,

$$K[P^{(o)}; \{p_j^{(o)}(1+z_j)\}_{j=1}^{n}] = \frac{1}{2} \sum_{j=1}^{n} p_j^{(o)} z_j^{\,2} \, . \tag{1.24}$$

We may use the properties of the contrast K to show that entropy is convex, i.e., that for two arbitrary different probability distributions $P^{(1)}$

and $P^{(2)}$

$$S(\beta P^{(1)} + (1-\beta)\, P^{(2)}) - \beta S(P^{(1)}) - (1-\beta) S(P^{(2)}) > 0; \tag{1.25}$$

where $0 < \beta < 1$. The left-hand side may be written

$$\beta \sum_i p_i^{(1)} \ln \frac{p_i^{(1)}}{\beta p_i^{(1)} + (1-\beta) p_i^{(2)}} + (1-\beta) \sum_i p_i^{(2)} \ln \frac{p_i^{(2)}}{\beta p_i^{(1)} + (1-\beta) p_i^{(2)}} =$$

$$= \beta\, K[\beta P^{(1)} + (1-\beta)\, P^{(2)};\, P^{(1)}] +$$
$$+ (1-\beta)\, K[\beta P^{(1)} + (1-\beta)\, P^{(2)};\, P^{(2)}],$$

and positivity then follows from (1.18) and (1.19).

Let us now assume that we have a system A composed of m subsystems, $A_1, \ldots, A_m$, and that the combined probability distribution for A is $P=\{p_I\}$, where $I = (i_1, \ldots, i_m)$ labels the states of the subsystems. If the subsystems are independent, then p_I factorizes. If this is not the case, then there are correlations between the subsystems. This contains an amount of information, often called the mutual information G of $A_1, \ldots, A_m$,

$$G[P] = K[\prod_{k=1}^{m} P^{(k)}; P] = \sum_{i_1, \ldots, i_m} p_{i_1 \ldots i_m} \ln \frac{p_{i_1 \ldots i_m}}{p_{i_1}^{(1)} \cdots p_{i_m}^{(m)}}\,;$$

$$P^{(k)} = \{p_i^{(k)}\}; \quad p_i^{(k)} = \sum_{i_1, \ldots, i_m} \delta_{i\, i_k}\, p_{i_1 \ldots i_m}\,. \tag{1.26}$$

1.2 The Maximum Entropy Principle and the Gibbs Distribution

The macrostate assignments above were done presuming that there were no constraints on the system that had to be taken into account. When a system is subject to certain constraints, which macrostate is then best suited to describe it? According to the *maximum entropy principle* [Jaynes 1963, Levine and Tribus 1979], it is the probability distribution which maximizes the entropy under those constraints. It has also been called the 'minimum bias principle', since it is equivalent to using only such bias in the choice of macrostate as is contained in the constraints.

The argument leading to the principle is straightforward. Let the constraints be of the kind (1.12), and let the maximum entropy value consistent with these constraints be S_{max}. Now if we choose a macrostate P (fulfilling the constraints) to describe A such that $S(P) < S_{max}$ then according to (1.14) there is a positive negentropy, $\Sigma(P) = S_{max} - S(P)$. This amount of information corresponds to further restrictions on the system, which must involve some ambiguously added information. The only way to avoid introducing such information is to choose P such that $S(P) = S_{max}$.

We shall in the following require that macrostates describing the system under consideration satisfy constraints (1.12) which are linear in the probabilities,

$$\langle f_k \rangle \equiv \sum_{j=1}^{n} p_j f_k^{(j)} = F_k \quad (k = 1, \dots, r), \tag{1.27}$$

where f_k are real functions of the microstates, which take on values $f_k^{(1)}, \dots, f_k^{(n)}$, and where '$< >$' denotes mean value. The problem is to find the probability distribution P which maximizes the entropy (1.10) subject to the normalization and positivity requirements (1.7) and (1.8) as well as the constraints (1.27). To solve this problem we look for an S which is stationary under variations of P

$$0 = -\,\mathrm{d}S = \sum_{j=1}^{n} \mathrm{d}p_j \,(\ln p_j + 1), \tag{1.28}$$

where the variations of P are limited under the constraints on $\mathrm{d}p_j$ corresponding to (1.8) and (1.27)

$$\begin{cases} \sum_{j=1}^{n} \mathrm{d}p_j = 0; & (1.29) \\ \sum_{j=1}^{n} \mathrm{d}p_j f_k^{(j)} = 0 \quad (k = 1, \ldots, r). & (1.30) \end{cases}$$

Combining (1.28) with (1.29) and (1.30) multiplied by (real) Lagrange multipliers $\mu - 1$ and y_k we get

$$\sum_{j=1}^{n} \mathrm{d}p_j \,(\ln p_j + \mu + \boldsymbol{y}\cdot\boldsymbol{f}^{(j)}) = 0, \tag{1.31}$$

where we have introduced a vector notation for the r-dimensional constraints,

$$\begin{cases} \boldsymbol{y} = (y_1, \ldots, y_r); \\ \boldsymbol{f}^{(j)} = (f_1^{(j)}, \ldots, f_r^{(j)}). \end{cases} \tag{1.32}$$

Now (1.31) is to hold for arbitrary $\mathrm{d}p_j$, and hence we have the *Gibbs distribution*,

$$p_j = \mathrm{e}^{-\mu - \boldsymbol{y}\cdot\boldsymbol{f}^{(j)}} . \tag{1.33}$$

The positivity condition (1.7) is then automatically satisfied. The normalization (1.8) determines μ as a function of $\boldsymbol{y}$,

$$\mu(\boldsymbol{y}) = \ln Z(\boldsymbol{y}), \tag{1.34}$$

where $Z(\boldsymbol{y})$ is the partition function

$$Z(\boldsymbol{y}) = \sum_{j=1}^{n} \mathrm{e}^{-\boldsymbol{y}\cdot\boldsymbol{f}^{(j)}} . \tag{1.35}$$

Finally $\boldsymbol{y}$ is determined by the constraints (1.27), most simply expressed as

$$\frac{\partial \mu(\boldsymbol{y})}{\partial \boldsymbol{y}} = -\boldsymbol{F} . \tag{1.36}$$

The maximum value of the entropy obtains for the probability distribution (1.33) and is

$$S = \mu + \boldsymbol{y}\cdot\boldsymbol{F} . \tag{1.37}$$

According to (1.36), the differentials of μ and S are

$$\begin{cases} \mathrm{d}\mu = -\boldsymbol{F}\cdot\mathrm{d}\boldsymbol{y} ; \\ \mathrm{d}S = \boldsymbol{y}\cdot\mathrm{d}\boldsymbol{F} . \end{cases} \tag{1.38}$$

Hence it is most convenient to view μ as a function of $\boldsymbol{y}$ (as we have done) and S as a function of $\boldsymbol{F}$. S and μ are so-called Legendre transforms of each other.

The matrices formed by the second partial derivatives

$$\begin{cases} A_{kl} = \dfrac{\partial^2 \mu}{\partial y_k \partial y_l} = -\dfrac{\partial F_k}{\partial y_l} = -\dfrac{\partial F_l}{\partial y_k} ; \\ B_{kl} = -\dfrac{\partial^2 S(\boldsymbol{F})}{\partial F_k \partial F_l} = -\dfrac{\partial y_k}{\partial F_l} = -\dfrac{\partial y_l}{\partial F_k} , \end{cases} \tag{1.39}$$

are clearly each others' inverses,

$$\mathbf{AB} = \mathbf{1}. \tag{1.40}$$

Furthermore A is the covariance matrix,

$$A_{kl} = < (f_k - F_k)(f_l - F_l) > = < f_k f_l > - F_k F_l . \tag{1.41}$$

To show this, we use (1.34), (1.35), (1.33) and (1.27):

$$A_{kl} = \frac{\partial^2 \mu}{\partial y_k \, \partial y_l} = -\frac{\partial}{\partial y_k} \frac{1}{Z} \sum_{j=1}^{n} f_l^{(j)} \, \mathrm{e}^{-\boldsymbol{y}\cdot\boldsymbol{f}^{(j)}} =$$

$$= -\frac{1}{Z^2} \sum_{j'=1}^{n} f_k^{(j')} \, \mathrm{e}^{-\boldsymbol{y}\cdot\boldsymbol{f}^{(j')}} \sum_{j=1}^{n} f_l^{(j)} \, \mathrm{e}^{-\boldsymbol{y}\cdot\boldsymbol{f}^{(j)}} + \frac{1}{Z} \sum_{j=1}^{n} f_k^{(j)} f_l^{(j)} \, \mathrm{e}^{-\boldsymbol{y}\cdot\boldsymbol{f}^{(j)}} =$$

$$= -\sum_{j,j'=1}^{n} p_j f_l^{(j)} p_{j'} f_k^{(j')} + \sum_{j=1}^{n} p_j f_k^{(j)} f_l^{(j)} = \langle f_k f_l \rangle - F_k F_l \, .$$

Then, for an arbitrary unit vector $\boldsymbol{e}$

$$\boldsymbol{e}\cdot(\mathbf{A}\boldsymbol{e}) = \langle (\boldsymbol{e}\cdot(\boldsymbol{f} - \boldsymbol{F}))^2 \rangle \geq 0. \tag{1.42}$$

We may assume strict inequality in (1.42), since if $\boldsymbol{e}\cdot\boldsymbol{f}^{(j)}$ were fixed at $\boldsymbol{e}\cdot\boldsymbol{F}^{(j)}$ for all j, it would be considered as a constant parameter and not a variable depending on the state j. Thus we would need to consider only the state functions $\boldsymbol{f} - (\boldsymbol{f}\cdot\boldsymbol{e})\boldsymbol{e}$ which span an $(r-1)$-dimensional ‘constraint space.’

Then the matrix $\mathbf{A}$ is positive definite, and because of (1.40) the same is true of $\mathbf{B}$. Hence (1.39) tells us that $\mu(\boldsymbol{y})$ is a convex function and that $S(\boldsymbol{F})$ is concave. The convexity of $\mu(\boldsymbol{y})$ ensures that $\boldsymbol{y}$ is uniquely determined by $\boldsymbol{F}$ through (1.36).

We note in passing that from $Z(\boldsymbol{y})$ or $\mu(\boldsymbol{y})$ also higher order moments may be similarly generated,

$$\langle f_{k_1}, \ldots, f_{k_m} \rangle = (-1)^m \, \mathrm{e}^{-\mu(\boldsymbol{y})} \frac{\partial}{\partial y_{k_1}} \cdots \frac{\partial}{\partial y_{k_m}} \mathrm{e}^{\mu(\boldsymbol{y})} \, . \tag{1.43}$$

1.2.1 *Contrast for two Gibbs distributions*

Very often one has reason to compare two maximum entropy states. For instance, we may compare the maximum entropy state of a subsystem to a similar state of the global system. Under constraints of the type (1.27) these states are described by Gibbs distributions.

We now consider two Gibbs distributions of the type (1.33),

$$\begin{cases} P = \{p_j\}_1^n; \quad p_j = e^{-\mu(y) - y \cdot f^{(j)}} ; \\ P^{(o)} = \{p_j^{(o)}\}_1^n; \quad p_j^{(o)} = e^{-\mu(y^{(o)}) - y^{(o)} \cdot f^{(j)}} , \end{cases} \tag{1.44}$$

with the following entropies and mean values of f,

$$\begin{cases} \sum_{j=1}^{n} p_j f^{(j)} = F ; \\ S = S[P] = \sum_{j=1}^{n} p_j \ln \frac{1}{p_j} = \mu(y) + y \cdot F , \end{cases} \tag{1.45}$$

and

$$\begin{cases} \sum_{j=1}^{n} p_j^{(o)} f^{(j)} = F^{(o)} ; \\ S^{(o)} = S[P^{(o)}] = \sum_{j=1}^{n} p_j^{(o)} \ln \frac{1}{p_j^{(o)}} = \mu(y^{(o)}) + y^{(o)} \cdot F. \end{cases} \tag{1.46}$$

Inserting (1.44) into (1.17) and using (1.45) and (1.46) we then obtain three different forms for K,

$$\begin{aligned} K[P^{(o)}; P] &= \mu(y^{(o)}) + y^{(o)} \cdot F - S = \\ &= y^{(o)} \cdot (F - F^{(o)}) - (S - S^{(o)}) = \\ &= \mu(y^{(o)}) - \mu(y) - \frac{\partial \mu(y)}{\partial y} \cdot (y^{(o)} - y). \end{aligned} \tag{1.47}$$

From the third expression here we conclude that if K is viewed as a function of $y^{(o)}$ and y

$$C(y^{(o)}; y) = K[P^{(o)}; P] \tag{1.48}$$

then (see (1.39))

$$\begin{cases} C(y^{(o)}; y) = 0; \\ \left(\dfrac{\partial C(y^{(o)}; y)}{\partial y}\right)_{y=y^{(o)}} = 0; \\ \left(\dfrac{\partial^2 C(y^{(o)}; y)}{\partial y_k\, \partial y_l}\right)_{y=y^{(o)}} = \dfrac{\partial^2 \mu(y^{(o)})}{\partial y_k^{(o)} \partial y_l^{(o)}} = A_{kl}(y^{(o)}). \end{cases} \tag{1.49}$$

Hence if $\mathbf{y}$ is close to $\mathbf{y}^{(o)}$ —which means that P is close to $P^{(o)}$ —then to second degree in $\mathbf{y} - \mathbf{y}^{(o)}$ we have

$$C(y^{(o)}; y) = \frac{1}{2}\,(y - y^{(o)})\cdot(\mathbf{A}(y - y^{(o)})). \tag{1.50}$$

This could also have been derived from (1.24), since z_j defined in (1.23) is, to first order in $\mathbf{y} - \mathbf{y}^{(o)}$,

$$z_j = -\,(\boldsymbol{f}^{(j)} - \boldsymbol{F}^{(o)})\cdot(y - y^{(o)}) \tag{1.51}$$

which inserted into (1.24) with use of (1.41) immediately leads to (1.50).

1.2.2 *Examples of discrete Gibbs distributions*

We shall illustrate the use of the maximum entropy principle and the Gibbs distribution in some well-known cases.

The Bose-Einstein distribution

Consider the simple case that the system is a set of individuals of some population, the state being described by the number n of individuals ($n = 0, 1, 2,\ldots$), and the only constraint being that the mean value of n is given as N. Then

$$\begin{cases} p_n \geq 0 \quad (n = 0, 1, 2, \ldots); \\ \sum_{n=0}^{\infty} p_n = 1; \\ \sum_{n=0}^{\infty} p_n n = N. \end{cases} \tag{1.52}$$

There is now only one parameter y. The partition function (1.35) is

$$e^{\mu(y)} = Z(y) = \sum_{n=0}^{\infty} e^{-yn} = \frac{1}{1 - e^{-y}}. \tag{1.53}$$

For this case (1.36) and (1.34) imply

$$N = -\frac{1}{Z}\frac{dZ}{dy} = \frac{1}{e^y - 1}, \tag{1.54}$$

and hence

$$\begin{cases} y = \ln\left(1 + \frac{1}{N}\right); \\ \mu = \ln \frac{e^y}{e^y - 1} = \ln(N+1); \\ S = (N+1)\ln(N+1) - N \ln N. \end{cases} \tag{1.55}$$

We recognize the probability distribution (1.33) as the well-known Bose-Einstein distribution

$$p_n = \frac{N^n}{(N+1)^{n+1}}. \tag{1.56}$$

The mean square variance is according to (1.39) and (1.41)

$$< (n - N)^2 > = \frac{d^2\mu}{dy^2} = \frac{e^y}{(e^y - 1)^2} = N(N+1). \tag{1.57}$$

The Bose-Einstein distribution is applicable to photons in incoherent radiation. We shall return to this in Section 2.2.3. It has also been applied to quantitative social science such as demography.

The Bernoulli distribution

Consider next a system consisting of n subsystems, each of them having two possible states, $m_j = 0, 1$ ($j = 1,\ldots, n$), altogether 2^n possible states. Let

$$m = \sum_{j=1}^{n} m_j \,, \tag{1.58}$$

and let the only constraint on the system be that

$$< m > = n\,c \,, \tag{1.59}$$

where c is a given number ($0 \le c \le 1$).

The probabilities are now $p_{m_1 \ldots m_n}$, and (1.59) may be written

$$\sum_{m_1 \ldots m_n = 0}^{1} p_{m_1 \ldots m_n}(m_1 + \ldots + m_n) = n\,c \,. \tag{1.60}$$

The partition function is now

$$Z(y) = \sum_{m_1 \ldots m_n = 0}^{1} \mathrm{e}^{-y(m_1 + \ldots + m_n)} = (1 + \mathrm{e}^{-y})^n \,, \tag{1.61}$$

and hence

$$\mu(y) = n \ln(1 + \mathrm{e}^{-y}). \tag{1.62}$$

Now (1.60) and (1.36) give

$$c = \frac{e^{-y}}{1+e^{-y}} = \frac{1}{1+e^{y}}. \tag{1.63}$$

and hence

$$\begin{cases} y = \ln \dfrac{1-c}{c}; \\ \mu = n \ln \dfrac{1}{1-c}; \\ S = n\left[c \ln \dfrac{1}{c} + (1-c) \ln \dfrac{1}{1-c} \right]. \end{cases} \tag{1.64}$$

The probability distribution is according to (1.33)

$$p_{m_1 \ldots m_n} = c^m (1-c)^{n-m}, \tag{1.65}$$

where m is given by (1.58). The probability distribution (1.65) is the Bernoulli distribution.

The binomial distribution

If the subsystems are indistinguishable then there is only one relevant parameter m, and the corresponding probability distribution is

$$p_m = \sum_{m_1 + \ldots + m_n = m} c^m (1-c)^{n-m} = \binom{n}{m} c^m (1-c)^{n-m} \quad (m = 1, \ldots, n), \tag{1.66}$$

i.e., the binomial probability distribution with $\langle m \rangle = nc$.

The Poisson distribution

If we put $nc = z$ in (1.66) and keep z finite while taking the limit $n \to \infty$, then using Stirling's formula ($n! \approx (2\pi)^{1/2} n^{n+1/2} e^{-n}$) we get

$$p_m = \mathrm{e}^{-z} \frac{z^m}{m!} \quad (m = 0, 1, 2, \ldots), \tag{1.67}$$

which is the Poisson distribution with the mean value $<m> = z$.

1.2.3 *Continuous state space*

Let us assume now that the space of state is the D-dimensional Euclidean space E^D, and let $\mathbf{x} = (x_1, \ldots, x_D)$ be the coordinate vector in this space. Let $\nu(\mathbf{x})$ be the density of states (the number of states per unit volume) at $\mathbf{x}$. Then corresponding to (1.8) we have for the non-negative probability distribution $p(\mathbf{x})$,

$$\int \mathrm{d}^D\mathbf{x}\ \nu(\mathbf{x})\, p(\mathbf{x}) = 1. \tag{1.68}$$

The constraints corresponding to (1.27) are

$$\int \mathrm{d}^D\mathbf{x}\ \nu(\mathbf{x})\, p(\mathbf{x})\, \boldsymbol{f}(\mathbf{x}) = \boldsymbol{F}\,. \tag{1.69}$$

The entropy is a functional of $p(\boldsymbol{x})$,

$$S[p] = \int \mathrm{d}^D\mathbf{x}\ \nu(\mathbf{x})\, p(\mathbf{x}) \ln \frac{1}{p(\mathbf{x})}\,. \tag{1.70}$$

Maximum entropy is found for the Gibbs distribution

$$p(\mathbf{x}) = \mathrm{e}^{-\mu(\boldsymbol{y}) - \boldsymbol{y}\cdot\boldsymbol{f}(\mathbf{x})}\,, \tag{1.71}$$

where

$$\begin{cases} \mu(\boldsymbol{y}) = \ln Z(\boldsymbol{y}); \\ Z(\boldsymbol{y}) = \int \mathrm{d}^D\mathbf{x}\ \nu(\mathbf{x})\, \mathrm{e}^{-\boldsymbol{y}\cdot\boldsymbol{f}(\mathbf{x})}\,. \end{cases} \tag{1.72}$$

Similarly to (1.36) and (1.37), $\boldsymbol{y}$ is determined by

$$\frac{\partial \mu(\boldsymbol{y})}{\partial \boldsymbol{y}} = -\boldsymbol{F}\,, \tag{1.73}$$

and the maximum value of the entropy is

$$S(\boldsymbol{F}) = \mu(y) + y \cdot \boldsymbol{F} \, . \tag{1.74}$$

The covariance matrix has similar properties as in the discrete case, and $\mu(y)$ is convex and $S(\boldsymbol{F})$ is concave.

The Gaussian distribution

Consider the case with a homogenous state density $\nu(\mathbf{x}) = 1$. Further let the constraints be

$$\begin{cases} <\mathbf{x}> = 0; \\ <\mathbf{x}\cdot\mathbf{x}> = \dfrac{D}{2b} \, . \end{cases} \tag{1.75}$$

where b is a positive number. Any condition fixing $<\mathbf{x}>$ and $<x_\alpha x_\beta>$ $(\alpha, \beta = 1,..., D)$ could be changed into (1.75) by a translation, a rotation, and change of scales along the axes.

The partition function (1.72) is then

$$Z(y) = \int \mathrm{d}^D\mathbf{x}\; \mathrm{e}^{-y\mathbf{x}^2} = \left(\frac{\pi}{y}\right)^{D/2}, \tag{1.76}$$

and

$$\mu(y) = \ln Z(y) = \frac{D}{2} \ln \frac{\pi}{y} \, , \tag{1.77}$$

where y is the variable connected to the second condition (1.75), and no other Lagrange multipliers appear in (1.76) or (1.77). Now (1.73) gives

$$y = b \, . \tag{1.78}$$

The probability distribution (1.71) is

$$p(\mathbf{x}) = \left(\frac{b}{\pi}\right)^{D/2} e^{-b\mathbf{x}^2}. \tag{1.79}$$

Hence also the Gaussian distribution is a special case of the Gibbs distribution. Its entropy, which may be obtained from (1.70) or (1.74), is

$$S = \frac{D}{2}\left(1 + \ln\frac{\pi}{b}\right) \tag{1.80}$$

The Gaussian distribution (1.79) and the binomial distribution (1.66) are the natural functions for describing broadening due to limited resolution in the continuous case and the discrete case, respectively. We shall use them for this purpose in Sections 3.1 and 3.3.

1.3 Rényi Entropies

Assume that the system of interest A consists of m similar subsystems, A_1, $A_2, \ldots, A_m$, each with the same a priori probability distribution P. A priori we also conjecture that the subsystems are uncorrelated. The macrosystem for A is thus described by a product probability distribution with the elements

$$p_{i_1 \ldots i_m} = p_{i_1} p_{i_2} \cdots p_{i_m}. \tag{1.81}$$

The probability that all subsystems are clustered in the same state, regardless which, is then

$$\sum_{i=1}^{n} p_{i \ldots i} = \sum_{i=1}^{n} p_i^{\,m}. \tag{1.82}$$

The information gained per extra subsystem if we learn that this is the case is

$$-\frac{1}{m-1}\ln\left(\sum_{i=1}^{n} p_i{}^m\right)=S_m[P], \tag{1.83}$$

where $S_m[P]$ is the *Rényi entropy* [Rényi 1970],

$$S_\alpha[P]=-\frac{1}{\alpha-1}\ln\left(\sum_{i=1}^{n} p_i{}^\alpha\right). \tag{1.84}$$

The Rényi entropies for integer $\alpha > 1$ are thus measures for the a priori clustering properties inherent in the single system distribution p_i. The definition (1.84) may be used even if α is not a positive integer. For negative α, S_α diverges unless all $p_i > 0$. For $\alpha \to 1$, $S_\alpha[P]$ is of the form 0/0, but it is easy to convince oneself that in this limit one simply obtains the ordinary entropy (1.10), i.e.,

$$S_1[P]=S[P]. \tag{1.85}$$

One can show that the Rényi entropies satisfy

$$\begin{cases} S_\alpha[P]\geq 0; \\ S_\alpha[\tilde{P}]=\ln n, \qquad \tilde{p}_1=\ldots=\tilde{p}_n=\dfrac{1}{n}, \end{cases} \tag{1.86}$$

and further

$$\begin{cases} S_\alpha[P]\geq \ln n\,, & \alpha<0; \\ S_0[P]=\ln n\,; \\ S_\alpha[P]\leq \ln n\,, & \alpha>0, \end{cases} \tag{1.87}$$

and additivity for uncorrelated probability distributions,

$$S_\alpha[PQ]=S_\alpha[P]+S_\alpha[Q]. \tag{1.88}$$

The Rényi entropy is convex only in a limited interval of α (cf. (1.25)),

$$S_\alpha[\beta P^{(1)}+(1-\beta)\,P^{(2)}]-\beta S_\alpha[P^{(1)}]-(1-\beta)S_\alpha[P^{(2)}]>0;\ 0<\alpha\leq 2, \tag{1.89}$$

As a function of α the Rényi entropy is monotonously decreasing [Hentschel and Procaccia 1983],

$$\frac{\partial}{\partial\alpha}S_\alpha[P] = -\frac{1}{(\alpha-1)^2}K[P; Z_\alpha] < 0; \quad Z_\alpha = \left\{ \frac{p_i^{\alpha}}{\sum_{k=1}^{n} p_k^{\alpha}} \right\}_{i=1}^{n} . \tag{1.90}$$

For infinite $|\alpha|$ the Rényi entropy is determined by the largest or smallest probability

$$\begin{cases} \lim_{\alpha\to+\infty} S_\alpha[P] = \ln\dfrac{1}{p_{\max}} , \\ \lim_{\alpha\to-\infty} S_\alpha[P] = \ln\dfrac{1}{p_{\min}} . \end{cases} \tag{1.91}$$

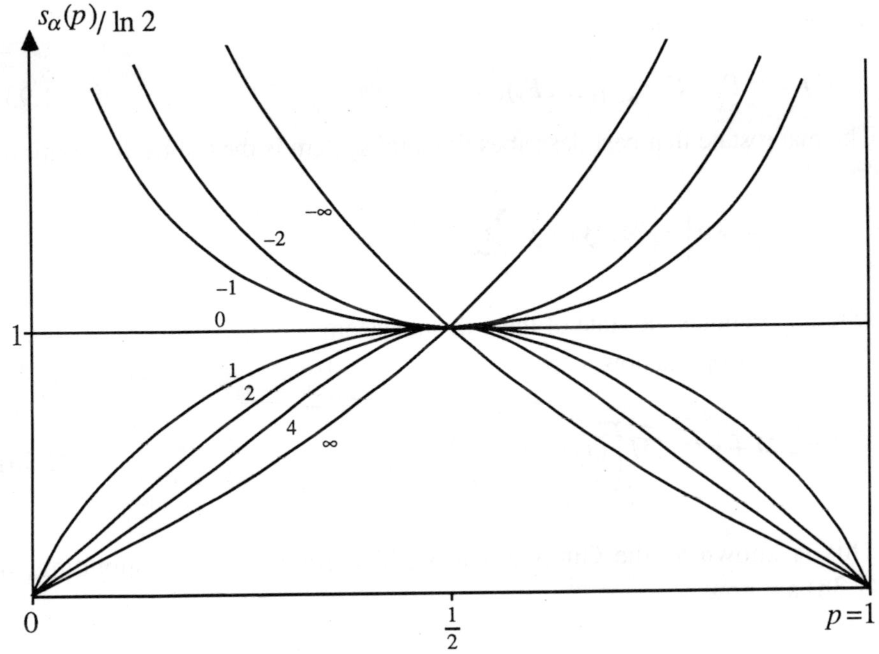

Fig. 1.2. Rényi entropies, $S_\alpha[\{p, 1-p\}] = s_\alpha(p)$, for varying α.

For $n = 2$ ($P = \{p, 1-p\}$) the properties of the Rényi entropy for varying α are illustrated in Fig. 1.2.

Rényi entropies have been used to describe clustering in dynamical systems, see e.g., [Grassberger and Procaccia 1982, Procaccia 1985].

1.4 Generalized Work and Heat

Consider now a system A consisting of M similar subsystems, assumed to be independent or to interact in a way which does not change the probability distribution [Brønsted 1955, Smith-Sørensen 1976, 1977]. We assume that we have additive variables

$$f = \sum_{m=1}^{M} f_m , \tag{1.92}$$

with given mean values

$$<f> = \boldsymbol{F}; \quad \boldsymbol{F} = (F_1, \ldots, F_r). \tag{1.93}$$

The macrostate that best describes the total system is the Gibbs distribution

$$p_{j_1 \ldots j_M} = \exp\left\{ -\left[M\mu(\boldsymbol{y}) + \boldsymbol{y}\cdot \sum_{m=1}^{M} \boldsymbol{f}^{(j_m)} \right] \right\}, \tag{1.94}$$

where j_m denotes the state of the mth subsystem.

The entropy is now

$$S = \mu M + \boldsymbol{y}\cdot\boldsymbol{F} = M\left(\mu(\boldsymbol{y}) - \boldsymbol{y}\cdot \frac{\partial \mu(\boldsymbol{y})}{\partial \boldsymbol{y}} \right) \tag{1.95}$$

This is known as the Gibbs equation. The equations corresponding to (1.38) are now

$$\begin{cases} M\,\mathrm{d}\mu + \boldsymbol{F}\cdot\mathrm{d}\boldsymbol{y} = 0; \\ \mathrm{d}S = \mu\,\mathrm{d}M + \boldsymbol{y}\cdot\mathrm{d}\boldsymbol{F}. \end{cases} \tag{1.96}$$

The second equation (1.96) is the Gibbs equation for differential quantities; the first equation (1.96) is known as the Gibbs-Duhem relation. It fixes the relation between $\boldsymbol{y}$ and $\boldsymbol{F}/M$, the mean value of $\boldsymbol{f}_m$, independent of m.

Now M and $\boldsymbol{F}$ are extensive variables with additive properties, and μ and $\boldsymbol{y}$ are the corresponding intensive variables.

The contrast corresponding to (1.47) and (1.48) is now

$$\begin{aligned} C(\boldsymbol{y}^{(0)}; \boldsymbol{y}) &= K[P^{(0)}; P] = \mu(\boldsymbol{y}^{(0)})\,M + \boldsymbol{y}^{(0)}\cdot\boldsymbol{F} - S = \\ &= \boldsymbol{y}^{(0)}\cdot(\boldsymbol{F} - \boldsymbol{F}^{(0)}) - (S - S^{(0)}) = \\ &= M\left[\mu(\boldsymbol{y}^{(0)}) - \mu(\boldsymbol{y}) - \frac{\partial\mu(\boldsymbol{y})}{\partial\boldsymbol{y}}\cdot(\boldsymbol{y}^{(0)} - \boldsymbol{y})\right]. \end{aligned} \tag{1.97}$$

The equations corresponding to (1.41) and (1.49) are now

$$< (f_k - F_k)(f_l - F_l) > = A_{kl} = \left(\frac{\partial^2 C(\boldsymbol{y}^{(0)}, \boldsymbol{y})}{\partial y_k\,\partial y_l}\right)_{\boldsymbol{y}=\boldsymbol{y}^{(0)}} = M\,\frac{\partial^2\mu(\boldsymbol{y}^{(0)})}{\partial y_k^{(0)}\partial y_l^{(0)}}, \tag{1.98}$$

which is the expression for fluctuations around a mean for an extended system.

1.4.1 *Work and heat*

We shall now consider the case where the state variables $\boldsymbol{f}^{(j)}$ depend on an external parameter L, along which work may be done [Jaynes 1963],

$$\boldsymbol{f}^{(j)} = \boldsymbol{f}^{(j)}(L). \tag{1.99}$$

Then also the partition function and its logarithm depend on L

$$\begin{cases} Z(\boldsymbol{y}, L) = \mathrm{e}^{M\mu(\boldsymbol{y},\, L)}; \\ \mu(\boldsymbol{y}, L) = \ln \sum_{j=1}^{n} \mathrm{e}^{-\boldsymbol{y}\cdot \boldsymbol{f}^{(j)}(L)}. \end{cases} \tag{1.100}$$

The probability distribution is (cf. (1.94))

$$p_{j_1 \cdots j_M}(L) = \exp\left\{ -\left[M\mu(\boldsymbol{y}, L) + \boldsymbol{y}\cdot \sum_{m=1}^{M} \boldsymbol{f}^{(j_m)}(L) \right] \right\}. \tag{1.101}$$

The mean values of $\boldsymbol{f}$ defined by (1.92) are (see (1.36))

$$\boldsymbol{F} = \langle \boldsymbol{f} \rangle = -M \frac{\partial \mu(\boldsymbol{y}, L)}{\partial \boldsymbol{y}}, \tag{1.102}$$

where $\boldsymbol{f}$ is defined by (1.92).

We can now study internal, i.e., macrostate changes in A, combined with external changes, i.e., motion along L. Differentiation of the second equation (1.100) gives

$$M\, \mathrm{d}\mu = -\boldsymbol{F}\cdot \mathrm{d}\boldsymbol{y} - \boldsymbol{y}\cdot \boldsymbol{x}\, \mathrm{d}L, \tag{1.103}$$

where

$$\boldsymbol{x} = \langle \frac{\partial \boldsymbol{f}}{\partial L} \rangle. \tag{1.104}$$

The entropy is

$$S(\boldsymbol{F}, L) = M\mu(\boldsymbol{y}, L) + \boldsymbol{y}\cdot \boldsymbol{F}. \tag{1.105}$$

Its differential is

$$\mathrm{d}S = \mu\, \mathrm{d}M + \boldsymbol{y}\cdot (\mathrm{d}\boldsymbol{F} - \boldsymbol{x}\, \mathrm{d}L). \tag{1.106}$$

Here (1.105) is the Gibbs equation as before. Its differential version (1.106) has changed as compared to (1.96), and so has the Gibbs-Duhem relation (1.103).

The Gibbs equation (1.106) implies

$$\Delta S = \mu\, \Delta M + \boldsymbol{y}\cdot\Delta \boldsymbol{Q}, \tag{1.107}$$

where we have introduced (small) 'heat differences'

$$\Delta \boldsymbol{Q} = \Delta \boldsymbol{F} - \boldsymbol{x}\, \Delta L\,. \tag{1.108}$$

It is then natural to regard

$$\Delta \boldsymbol{W} = \boldsymbol{x}\, \Delta L \tag{1.109}$$

as 'work' and hence to regard $\boldsymbol{x}$ as 'forces.' Hence $\Delta \boldsymbol{F}$ consists of a heat part and a work part,

$$\Delta \boldsymbol{F} = \Delta \boldsymbol{Q} + \Delta \boldsymbol{W}. \tag{1.110}$$

As (1.107) shows, only exchange of heat leads to a change in entropy.

1.4.2 *Work cycles and exergy*

We consider a system A (Fig. 1.3), with thermodynamical variables $\boldsymbol{F}$ and $\boldsymbol{y}$, assumed to be in internal equilibrium during a work cycle. We assume that A has an exchange of conserved extensive quantities with a 'work reservoir' B with zero entropy and characterized by a unit vector $\boldsymbol{e}$, and

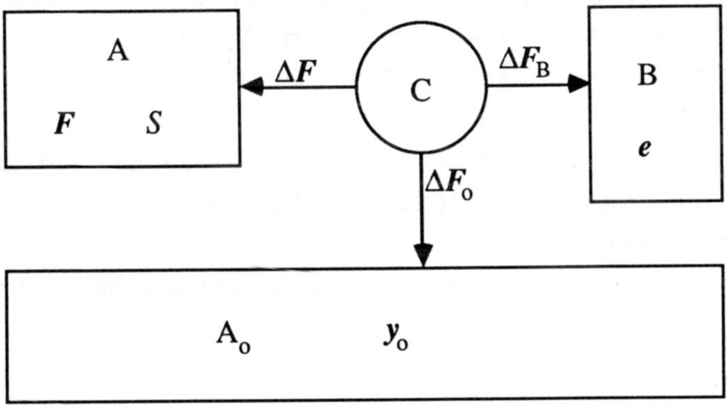

Fig. 1.3. The system A, exchanging work with the work reservoir B and heat with the heat reservoir A_o through the converter C.

with a 'heat reservoir' A_o, characterized by constant intensive variables $\boldsymbol{y}_o$. The exchange is mediated by the converter C. The converter is assumed to return to its original state after a full cycle. If ΔW is the work extracted (in fact, it is sufficient to assume that A consists of subsystems, such that each subsystem is in internal equilibrium, and we shall use such an assumption in Section 2.4.1), and $\Delta \boldsymbol{F}$ is the exchange with A_o, then conservation implies

$$\Delta \boldsymbol{F} + \Delta \boldsymbol{F}_B + \Delta \boldsymbol{F}_o = -\Delta \boldsymbol{F}_C = 0; \quad \Delta \boldsymbol{F}_B = \boldsymbol{e}\cdot\Delta W \tag{1.111}$$

The total entropy increase is non-negative,

$$\begin{aligned} 0 \le \Delta S_{tot} &= \Delta S + \Delta S_o = \Delta S + \boldsymbol{y}_o\cdot\Delta \boldsymbol{F}_o = \\ &= -\Delta(\boldsymbol{y}_o\cdot\boldsymbol{F} - S) - \boldsymbol{e}\cdot\boldsymbol{y}_o\,\Delta W. \end{aligned} \tag{1.112}$$

This may be expressed as

$$\Delta E + \Delta W = \frac{1}{\boldsymbol{e}\cdot\boldsymbol{y}_o}\,\Delta S_{tot} \le 0 \tag{1.113}$$

where (see (1.97))

$$E = \frac{1}{\boldsymbol{e}\cdot\boldsymbol{y}_o}\,(M\,\mu(\boldsymbol{y}_o) + \boldsymbol{y}_o\cdot\boldsymbol{F} - S) = \frac{1}{\boldsymbol{e}\cdot\boldsymbol{y}_o}\,C(\boldsymbol{y}_o;\,\boldsymbol{y}) \tag{1.114}$$

is the (generalized) *exergy* of A, i.e., the (generalized) work available in an ideal (reversible) exchange with A_o and B. As seen in (1.113) entropy production introduces losses in the exchange of work with B.

Entropy is at the same time an information-theoretical quantity and a physical quantity. What has been achieved in (1.113) and (1.114) is that the essentially physical quanities 'work' and 'exergy' have been tied to the information-theoretical quantity contrast [Jaynes 1963].

Exergy as a thermodynamical concept is the main theme of the next chapter.

2. EXERGETICS

Classical thermodynamics demonstrates that when a closed homogenous system changes isothermally and isobarically from one state to another, the maximal mechanical work output is the difference between the initial and final values of the total *Gibbs free energy*, $G=U+pV-TS$, of the system (U=internal energy, p=pressure, V=volume; S=entropy, T=temperature). Similarly, for an isothermal change we have the *Helmholtz free energy*, $F=U-TS$. Both free energies are *state functions*, because U, H, T, and S depend only on the state of the system, and they represent the upper limits to the mechanical work obtainable in certain transformations. It is then natural to ask for a more general measure of the maximal work yield of a system, applicable to general processes involving interactions, either internally or with other systems, as well as constraints. Such a general measure is provided through the exergy concept, formally introduced in Chapter 1. When work is extracted from an otherwise closed system, the exergy is a state function in the usual sense, depending only on variables of the system. If the system is brought to equilibrium with an external reference system, and work is extracted during this process, then the exergy also depends on the intensive variables of the reference system.

Various terms have been used to denote exergy: 'available energy', 'availability', 'énergie utilisable', and 'technische Arbeitsfähigkeit' are a few examples. These terms were often used to denote the maximal work in particular processes. The term 'exergy' was introduced by Rant [1956]. The concept can be found early in the history of thermodynamics:

> 'For example, let it be required to find the greatest amount of work which can be obtained from a given quantity of a certain substance in a given initial state, without increasing its total volume or allowing heat to pass to or from external bodies, except such as at the close of the processes are left in their initial condition. This has been called the *available energy* of the body.'
> Gibbs [1873] p. 49.

> 'The whole of this part of the subject [Thermodynamics] depends on the consideration of the Intrinsic Energy of a system of bodies, as depending on the temperature and physical state, as well as the form, motion, and relative position of these bodies. Of this energy, however, only a part is available for the purpose of producing mechanical work, and though the energy itself is indestructible, the available part is liable to diminuition by the action of certain

natural processes, such as conduction and radiation of heat, friction, and viscosity.'

Maxwell [1875] preface.

The historical development has gradually gone from the analysis of maximal useful work in particular processes and machines, present already in the work of Carnot [1824], to the more general framework adopted today. The foundation for such a generalization was provided when thermodynamics was integrated with statistical mechanics. Later, an important step was made by Jaynes [1963] when he connected statistical mechanics with information theory (see also Brønsted [1955], Tribus et al. [1966], and Levine and Tribus [1979]). The mechanical work view is still important in many applications [Ahern 1980, Gaggioli 1980, 1983, Moran 1982, Kotas 1985].

2.1 Thermodynamics

The presentation of information theory so far, has been modelled after statistical mechanics, the microscopic theory of thermodynamics. We shall now go in the other direction and consider statistical mechanics—and hence thermodynamics—(see any standard text-book in statistical mechanics, such as Reif [1965]) as an application of information theory [Jaynes 1963]. We first note that for entropy the unit systems differ by a factor of k (=Boltzmann's constant):

$$S(\text{thermodynamics}) = k\,S(\text{information theory}). \tag{2.1}$$

We further interprete the extension M used in Section 1.4 as related to the volume V. The corresponding intensive quantity $\mu(y)$ is then related to pressure. We thus make the substitution

$$\begin{cases} M \rightarrow V\,, \\ \mu(y) \rightarrow \dfrac{p}{k\,T}\,, \end{cases} \tag{2.2}$$

where p is pressure and T is temperature. Further, introducing the internal energy U, and the numbers of molecules N_i of the M chemical species ($i=1,\ldots, M$), with their chemical potentials μ_i (per molecule), we can interprete $\boldsymbol{F}$ and $\boldsymbol{y}$ according to

$$\begin{cases} \boldsymbol{F} = (U, N_1, \ldots, N_M); \\ \\ \boldsymbol{y} = \dfrac{1}{kT}\,(1, -\mu_1, \ldots, -\mu_M). \end{cases} \tag{2.3}$$

Work is usually done in the 'direction' of energy,

$$\boldsymbol{e} = (1, 0, 0, \ldots); \qquad \boldsymbol{e}\cdot\boldsymbol{y} = \frac{1}{kT}\,. \tag{2.4}$$

Then the generalized Gibbs relations and the Gibbs-Duhem relation, (1.95) and (1.96), take their ordinary form

$$\begin{cases} U = TS - pV + \sum_i \mu_i N_i\,; \\ \\ dU = TdS - pdV + \sum_i \mu_i\, dN_i\,, \end{cases} \tag{2.5}$$

$$SdT - Vdp + \sum_i N_i\, d\mu_i = 0, \tag{2.6}$$

and the exergy (1.114) is

$$\begin{aligned} E = k\,T_o\,K &= \left(U + p_o V - T_o S - \sum_i \mu_{io} N_i\right) = \\ &= U - U_{eq} + p_o(V - V_{eq}) - T_o(S - S_{eq}) - \sum_i \mu_{io}(N_i - N_{i\,eq}) = \\ &= S(T - T_o) - V(p - p_o) + \sum_i N_i\,(\mu_i - \mu_{io})\,, \end{aligned} \tag{2.7}$$

where T_o, p_o, μ_{io} are the intensive variables of the reference state (assumed to be in equilibrium), and where U_{eq}, S_{eq}, V_{eq}, and $N_{i\,eq}$ are the corresponding extensive quantities in the reference state of the studied system.

2.1.1 *Exergy as a thermodynamical state variable*

We shall now consider a system A, consisting of subsystems A_l (l=1,..., L), each in internal equilibrium (Fig. 2.1). Through exchange via a converter C the subsystems may exchange energy and chemical components while they deliver work to a work reservoir B. All processes are assumed to be ideal (reversible). We shall assume that A_l occupies a fixed space region of volume V_l. Originally the extensive variables are assumed to be U_l, S_l, V_l and N_{il}, and

$$U = \sum_{l=1}^{L} U_l \ ; \ S = \sum_{l=1}^{L} S_l \ ; \ V = \sum_{l=1}^{L} V_l \ ; \ N_i = \sum_{l=1}^{L} N_{il} \ . \tag{2.8}$$

The task is to find the exergy in A, i.e., the maximum amount of work that can be delivered to B in connection with internal exchanges and chemical reactions in A.

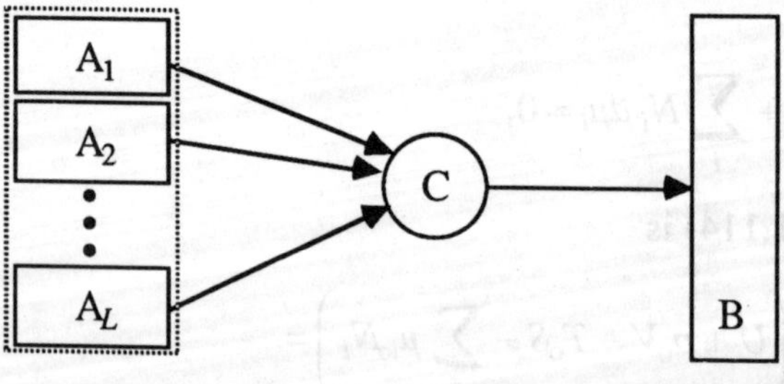

Fig. 2.1. A system A consisting of L subsystems A_1,..., A_L, delivering mechanical work to the work reservoir B through reversible processes via the converter C.

We use the circumflex mark to denote quantities connected to the final state, which has to be an equilibrium state for A. Then the exergy is determined by

$$E = U + \hat{p}V - \hat{T}S - \sum_{i=1}^{M} \hat{\mu}_i N_i \,. \tag{2.9}$$

The intensive quantities of the equilibrium state have to be determined from reversibility, from conservation laws such as stoichiometric constraints, and from the conditions for chemical equilibrium,

$$\begin{cases} \hat{S} = S\,; \\ \hat{V} = V \qquad \text{(in fact, } \hat{V}_l = V_l\text{);} \\ \hat{N}_i = N_i + \sum_{r=1}^{J} \nu_{ir}\,\xi_r \qquad (i = 1,\ldots, M); \\ \sum_{i=1}^{M} \nu_{ir}\hat{\mu}_i = 0 \qquad (r = 1,\ldots, J). \end{cases} \tag{2.10}$$

Here J is the number of chemical reactions that have to be taken into account; ν_{ir} is the stoichiometric coefficient of i in reaction r and ξ_r is the corresponding extent of reaction. In terms of the previous notation we have (see (1.36), (1.95), (2.3), and (2.4))

$$E = \boldsymbol{e}\cdot(\boldsymbol{F} - \hat{\boldsymbol{F}}), \tag{2.11}$$

and

$$\begin{cases} \mu(\hat{\boldsymbol{y}}) - \hat{\boldsymbol{y}}\cdot\dfrac{\partial\mu(\hat{\boldsymbol{y}})}{\partial\hat{\boldsymbol{y}}} = \dfrac{S}{N}\,; \\ \dfrac{\partial\mu(\hat{\boldsymbol{y}})}{\partial\hat{y}_{j+1}} = F_{j+1} + \sum_{r=1}^{J} \nu_{jr}\xi_r \qquad (j = 1,\ldots, M); \\ \sum_{i=1}^{M} \nu_{ir}\hat{y}_{i+1} = 0\,; \qquad (r = 1,\ldots, J). \end{cases} \tag{2.12}$$

These are $M + J + 1$ equations which uniquely determine $\xi_1,\ldots, \xi_J$ and $\hat{\boldsymbol{y}}$, due to the convexity of $\mu(y)$. In its turn $\hat{y}$ determines $\hat{F}$. Hence, the exergy E of the initial state is uniquely determined.

2.1.2 *Local thermodynamics*

To study transformations of energy and materials locally, the methods of hydrodynamics are readily available. Glansdorff and Prigogine [1971] combined hydrodynamics with thermodynamics under the assumption of local thermodynamic equilibrium. We shall follow their method quite closely. We then consider a system of several chemical species of molecular mass m_i and $b_i(\mathbf{x},t)$ molecules per unit mass. The mass density of each component is

$$\rho_i(\mathbf{x},t) = m_i\, b_i(\mathbf{x},t)\, \rho(\mathbf{x},t) \tag{2.13}$$

where $\rho(\mathbf{x},t)$ is the total mass density. Thus,

$$\rho(\mathbf{x},t) = \sum_i \rho_i(\mathbf{x},t). \tag{2.14}$$

For each species we introduce a velocity field

$$\mathbf{v}_i(\mathbf{x},t) = \mathbf{v}(\mathbf{x},t) + \mathbf{u}_i(\mathbf{x},t), \tag{2.15}$$

where $\mathbf{v}(\mathbf{x},t)$ is the center-of-mass velocity, and $\mathbf{u}_i(\mathbf{x},t)$ is the velocity of i in the local center-of-mass frame. Then, by definition,

$$\left\{\begin{array}{ll} \rho\mathbf{v} = \sum_i \rho_i \mathbf{v}_i\,, & (2.16) \\ \sum_i \rho_i u_i = 0. & (2.17) \end{array}\right.$$

The hydrodynamic time-derivative (i.e., for a volume element along its center-of-mass motion) is

$$\frac{\mathrm{d}}{\mathrm{d}t} = \frac{\partial}{\partial t} + \mathbf{v}\cdot\nabla \,. \tag{2.18}$$

Conservation of mass implies the continuity equation,

$$\frac{\partial \rho}{\partial t} + \nabla\cdot(\rho\mathbf{v}) = 0. \tag{2.19}$$

Let a be the amount per unit mass of some extensive quantity such that ρa is the corresponding density and $\rho a\mathbf{v}$ the corresponding current density. Then, because of (2.19), the balance equation for this quantity is

$$\frac{\partial(\rho a)}{\partial t} + \nabla\cdot(\rho a\mathbf{v}) = \rho\frac{\mathrm{d}a}{\mathrm{d}t} \,. \tag{2.20}$$

For the special choice $a = v$, where $v = \rho^{-1}$ is the specific volume, we find

$$\frac{\mathrm{d}v}{\mathrm{d}t} = v\,\nabla\cdot\mathbf{v} \,. \tag{2.21}$$

We have the stoichiometric relations

$$\sum_i m_i \nu_{ir} = 0 \qquad (r = 1,\ldots, J). \tag{2.22}$$

We also have a balance relation of the type (2.20) for each chemical component,

$$\frac{\partial \rho_i}{\partial t} + \nabla\cdot(\rho_i\mathbf{v}_i) = m_i \sum_r \nu_{ir}\,\xi_r \,. \tag{2.23}$$

The different molecular species may be subject to different external forces. Let the external force per unit mass for the ith component be $\mathbf{f}_i(\mathbf{x},t)$ and let the corresponding potential energy per unit mass be $\varphi_i(\mathbf{x},t)$. The force per unit mass of the mixture $\mathbf{f}(\mathbf{x},t)$ is determined by

$$\rho\,\mathbf{f} = \sum_i \rho_i \mathbf{f}_i \,, \tag{2.24}$$

and similarly for the total potential energy per unit mass φ,

$$\rho\varphi = \sum_i \rho_i \varphi_i \,. \tag{2.25}$$

There are also internal forces described by the pressure tensor, $\mathbf{P}$, which we assume to be of the form, valid for a fluid,

$$P_{\alpha\beta} = p\,\delta_{\alpha\beta} + P^{(d)}_{\alpha\beta} \qquad (\alpha,\beta = 1, 2, 3), \tag{2.26}$$

where p is the hydrostatic pressure and where

$$\begin{cases} P^{(d)}_{\alpha\beta} = P^{(d)}_{\beta\alpha} \qquad (\alpha,\beta = 1, 2, 3), \\ \displaystyle\sum_{\beta=1}^{3} P^{(d)}_{\beta\beta} = 0 \,. \end{cases} \tag{2.27}$$

The first term on the right-hand side in (2.26) is the elastic term whereas the second term is a dissipative term leading to entropy production. At equilibrium therefore $\mathbf{P}^{(d)} = 0$.

The equation of motion for a volume element is the balance equation (2.20) for momentum

$$\rho\,\frac{d\mathbf{v}}{dt} = \rho\,\mathbf{f} - \nabla\cdot\mathbf{P} \,. \tag{2.28}$$

For the time derivative of the kinetic energy density we get

$$\frac{\partial}{\partial t}\left(\frac{1}{2}\rho\mathbf{v}^2\right) = -\nabla\cdot\left(\frac{1}{2}\rho\mathbf{v}^2\mathbf{v}\right) - \mathbf{v}\cdot(\nabla\cdot\mathbf{P}) + \rho\mathbf{v}\cdot\mathbf{f}\,. \tag{2.29}$$

The kinetic energy current density is

$$\mathbf{P}\cdot\mathbf{v} + \frac{1}{2}\rho\mathbf{v}^2\mathbf{v}\ , \tag{2.30}$$

and the source of kinetic energy is thus

$$\frac{\partial}{\partial t}\left(\frac{1}{2}\rho\mathbf{v}^2\right) + \nabla\cdot\left(\mathbf{P}\cdot\mathbf{v} + \frac{1}{2}\rho\mathbf{v}^2\mathbf{v}\right) = \mathbf{P}:(\nabla\mathbf{v}) + \rho\mathbf{v}\cdot\mathbf{f}\,. \tag{2.31}$$

For the potential energy of each chemical component we have the balance equation

$$\frac{\partial}{\partial t}(\rho_i\varphi_i) + \nabla\cdot(\rho_i\varphi_i\mathbf{v}_i) = -\rho_i\,\mathbf{f}_i\cdot\mathbf{v}_i\,, \tag{2.32}$$

with the sum

$$\frac{\partial}{\partial t}(\rho\varphi) + \nabla\cdot\left(\sum_i \rho_i\varphi_i\mathbf{v}_i\right) = -\rho\,\mathbf{f}\cdot\mathbf{v} - \sum_i \rho_i\,\mathbf{f}_i\cdot\mathbf{u}_i\,. \tag{2.33}$$

Here we have used (2.24) and (2.25). For the sum of (2.31) and (2.33) we get

$$\frac{\partial}{\partial t}\left(\rho\left(\frac{1}{2}\mathbf{v}^2 + \varphi\right)\right) + \nabla\cdot\left(\mathbf{P}\cdot\mathbf{v} + \frac{1}{2}\rho\mathbf{v}^2\mathbf{v} + \sum_i \rho_i\varphi_i\mathbf{v}_i\right) =$$

$$= \mathbf{P}:(\nabla\mathbf{v}) - \sum_i \rho_i\,\mathbf{f}_i\cdot\mathbf{u}_i\,. \tag{2.34}$$

Besides kinetic and potential energy we have an internal energy density ρu and a corresponding current density

$$\rho u \mathbf{v} + \mathbf{q} \tag{2.35}$$

where the first term is the convective heat flow and the second term is the conductive heat flow. The continuity equation expressing overall energy conservation is then

$$\frac{\partial}{\partial t}\left[\rho\left(\frac{1}{2}\mathbf{v}^2 + \varphi + u\right)\right] + \\ + \nabla\cdot\left(\mathbf{P}\cdot\mathbf{v} + \frac{1}{2}\rho\mathbf{v}^2\mathbf{v} + \sum_i \rho_i\varphi_i\mathbf{v}_i + \rho u\mathbf{v} + \mathbf{q}\right) = 0 \tag{2.36}$$

Subtracting (2.34) from this we obtain for the internal energy the balance equation

$$\frac{\partial}{\partial t}(\rho u) + \nabla\cdot(\rho u\mathbf{v} + \mathbf{q}) = -\mathbf{P}:(\nabla\mathbf{v}) + \sum_i \rho_i \, \mathbf{f}_i\cdot\mathbf{u}_i \, . \tag{2.37}$$

To allow a complete thermodynamic analysis, we assume *local thermodynamic equilibrium*, i.e., validity of the local forms of the Gibbs and Gibbs-Duhem relations (2.5),

$$\begin{cases} u = Ts - pv + \sum_i \mu_i b_i \\ du = T\,ds - p\,dv + \sum_i \mu_i\,db_i \, , \end{cases} \tag{2.38}$$

which allows us to introduce the entropy density ρs. Along the centre-of-mass motion one has

$$\frac{ds}{dt} = \frac{1}{T}\frac{du}{dt} + \frac{p}{T}\frac{dv}{dt} - \sum_i \frac{\mu_i}{T}\frac{db_i}{dt} \, . \tag{2.39}$$

We have already by (2.37), (2.21), and (2.23)

$$\begin{cases} \rho \dfrac{\mathrm{d}u}{\mathrm{d}t} = -\nabla\cdot\mathbf{q} - \mathbf{P}:(\nabla\mathbf{v}) + \sum_i \rho_i\, \mathbf{f}_i\cdot\mathbf{u}_i \;; \\ \rho \dfrac{\mathrm{d}v}{\mathrm{d}t} = \nabla\cdot\mathbf{v}\;; \\ \rho \dfrac{\mathrm{d}b_i}{\mathrm{d}t} = \dfrac{1}{m_i}\left(\dfrac{\partial\rho_i}{\partial t} + \nabla\cdot(\rho_i\mathbf{v}_i) - \nabla\cdot(\rho_i\mathbf{u}_i)\right) = \sum_r \nu_{ir}\,\xi_r - \nabla\cdot(\rho\, b_i\mathbf{u}_i), \end{cases} \tag{2.40}$$

which inserted into (2.39) yields

$$\frac{\partial(\rho s)}{\partial t} + \nabla\cdot(\rho s\mathbf{v}) = -\frac{1}{T}\nabla\cdot\mathbf{q} - \frac{1}{T}\,\mathbf{P}^{(\mathrm{d})}:(\nabla\mathbf{v}) + \sum_i \frac{\mu_i}{T}\nabla\cdot(\rho\, b_i\mathbf{u}_i) + \\ + \sum_r \frac{A_r}{T}\,\xi_r + \frac{1}{T}\sum_i \rho_i\, \mathbf{f}_i\cdot\mathbf{u}_i \;. \tag{2.41}$$

Here A_r is the affinity for the rth reaction,

$$A_r = -\sum_i \nu_{ir}\,\mu_i \qquad (r = 1,\ldots, J). \tag{2.42}$$

We may consider (2.41) as a balance equation for entropy,

$$\frac{\partial(\rho s)}{\partial t} + \nabla\cdot\mathbf{j}_\mathrm{s} = \sigma\,, \tag{2.43}$$

where

$$\mathbf{j}_\mathrm{s} = \rho s\mathbf{v} + \frac{1}{T}\mathbf{q} - \rho\sum_i \frac{\mu_i}{T}\, b_i\mathbf{u}_i \tag{2.44}$$

is the entropy current density and

$$\sigma = \mathbf{q}\cdot\nabla\left(\frac{1}{T}\right) - \rho\sum_i b_i\mathbf{u}_i\cdot\left[\nabla\left(\frac{\mu_i}{T}\right) - \frac{m_i}{T}\,\mathbf{f}_i\right] - \\ - \frac{1}{T}\,\mathbf{P}^{(\mathrm{d})}:(\nabla\mathbf{v}) + \sum_r \frac{A_r}{T}\,\xi_r \tag{2.45}$$

is the entropy production density. The four terms in (2.45) are of the form 'flow' times 'force.' They represent heat conduction, diffusion, internal friction, and chemical reactions.

The exergy density is in analogy to (2.7)

$$\rho e = \rho\left(\frac{1}{2}\mathbf{v}^2 + \varphi + u\right) + p_\mathrm{o} - \rho\left(T_\mathrm{o}s + \sum_i b_i\mu_\mathrm{io}\right), \tag{2.46}$$

where e is the exergy per unit mass. With (2.38) and (2.25),

$$e = \frac{1}{2}\mathbf{v}^2 - v\,(p - p_\mathrm{o}) + s\,(T - T_\mathrm{o}) + \sum_i b_i\,(m_i\varphi_i + \mu_i - \mu_\mathrm{io}). \tag{2.47}$$

The exergy current density is similarly defined as (see (2.36) and (2.44))

$$\mathbf{j}_\mathrm{e} = \mathbf{P}\cdot\mathbf{v} + \frac{1}{2}\rho\,\mathbf{v}^2\,\mathbf{v} + \sum_i \mathbf{v}_i\rho_i\varphi_i + \rho u\mathbf{v} + \mathbf{q} - T_\mathrm{o}\mathbf{j}_\mathrm{s} - \sum_i \mathbf{v}_i\rho b_i\mu_\mathrm{io}. \tag{2.48}$$

Since the volume term of the exergy density is constant in (2.46), we have excluded the corresponding flow here. We can rewrite (2.48) in a way similar to (2.47) as

$$\mathbf{j}_\mathrm{e} = (\rho e + p - p_\mathrm{o})\mathbf{v} + \mathbf{P}^{(\mathrm{d})}\cdot\mathbf{v} + $$

$$+T_\mathrm{o}\left[\mathbf{q}\left(\frac{1}{T_\mathrm{o}} - \frac{1}{T}\right) + \rho\sum_i b_i\mathbf{u}_i\left(\frac{\mu_i}{T} - \frac{\mu_\mathrm{io}}{T_\mathrm{o}} + \frac{m_i\varphi_i}{T_\mathrm{o}}\right)\right]. \tag{2.49}$$

The exergy consumption density is then

$$\varepsilon = -\left[\frac{\partial(\rho e)}{\partial t} + \nabla\cdot\mathbf{j}_\mathrm{e}\right] = T_\mathrm{o}\,\sigma\,. \tag{2.50}$$

We have assumed that the reference state is in chemical equilibrium, i.e., that

$$\sum_i \nu_{ir}\,\mu_\mathrm{io} = 0 \qquad (r = 1,\ldots, J). \tag{2.51}$$

2.1.3 *Evolution towards equilibrium or a stationary state*

As in the previous section we shall assume local thermodynamic equilibrium. Moreover we shall assume that variations of densities, temperature, etc., within the system are limited so that certain linear relations may be assumed. As we shall see, if such a system has stationary boundary conditions it evolves towards a steady state characterized by minimum entropy production.

We shall let our system be characterized by the variables

$$\alpha_i(\mathbf{x},t) = \frac{\zeta_i(\mathbf{x},t)}{\zeta_{i\,\mathrm{eq}}} - 1 \qquad (i = 0, 1, 2, \ldots), \tag{2.52}$$

where ζ_β are the densities; $\zeta_o = \rho u$, and where $\zeta_{o\,\mathrm{eq}} = (\rho u)_{\mathrm{eq}}$ and $\zeta_{\beta\,\mathrm{eq}}$ (β=1, 2,...) are equilibrium values. The corresponding currents ('flows') are

$$\mathbf{J}_i(\mathbf{x},t) = \frac{\mathbf{j}_i(\mathbf{x},t)}{\zeta_{i\,\mathrm{eq}}} \qquad (i = 0, 1, 2, \ldots). \tag{2.53}$$

We shall consider the case that there is no net center-of-mass flow, nor any external force, and no chemical reaction, i.e., $\mathbf{v} = 0$ and all $\xi_r = 0$. Then (2.37) and (2.23) simplify,

$$\dot{\alpha}_i + \nabla\cdot\mathbf{J}_i = 0\,. \tag{2.54}$$

Within the approximations we shall use, the negentropy of our system is (summation over repeated indices is understood)

$$\Sigma(t) = S_{\max} - S(t) = \int_V \mathrm{d}^3\mathbf{x} \left\{ \tfrac{1}{2}\, \alpha_i(\mathbf{x},t) G_{ij} \alpha_j(\mathbf{x},t) \right\} \geq 0\,, \tag{2.55}$$

where $\{G_{ij}\}$ is a symmetric positive definite matrix. Inhomogeneities are characterized by 'forces,'

$$\mathbf{X}_i\,(\mathbf{x},t) = -\nabla \frac{\delta\Sigma}{\delta\,\alpha_i(\mathbf{x},t)} = -\,G_{ij}\,\nabla\alpha_j(\mathbf{x},t), \tag{2.56}$$

which 'drive' the system towards equilibrium.

If we are close enough to equilibrium we may assume linear relations between flows and forces

$$\mathbf{J}_i = \mathbf{L}_{ij}\cdot\mathbf{X}_j = -\,(\mathbf{L}_{ik}G_{kj})\cdot\nabla\alpha_j\,, \tag{2.57}$$

where Onsager's relations

$$\mathrm{L}_{i\gamma j\beta} = \mathrm{L}_{j\beta i\gamma} \tag{2.58}$$

hold, expressing the principle of detailed balancing, which may be derived from microscopic invariance under time reversal. Moreover $\mathbf{L}_{ij}$ is positive definite.

We can now study the time development of $\alpha_i(\mathbf{x},t)$. We get from (2.54) and (2.57)

$$\dot{\alpha}_i(\mathbf{x},t) = -\nabla\cdot\mathbf{J}_i\,(\mathbf{x},t) = \nabla\cdot((\mathbf{L}_{ik}G_{kj})\cdot\nabla\alpha_j(\mathbf{x},t)), \tag{2.59}$$

which leads to the following time development

$$\begin{cases} \alpha_i(\mathbf{x}\,,t) = (2\pi)^{-3/2} \displaystyle\int \mathrm{d}^3\mathbf{p}\; \mathrm{e}^{i\,\mathbf{p}\cdot\mathbf{x}}\,(\mathrm{e}^{-\mathbf{p}\cdot\mathrm{L}\cdot\mathbf{p}\;Gt})_{ij}\,\tilde{\alpha}_j\,(\mathbf{p})\,; \\ \tilde{\alpha}_j(\mathbf{p}) = (2\pi)^{-3/2} \displaystyle\int \mathrm{d}^3\mathbf{x}\; \mathrm{e}^{i\,\mathbf{p}\cdot\mathbf{x}}\;\alpha_j(\mathbf{x},0)\,, \end{cases} \tag{2.60}$$

i.e., an exponential decay of the deviation from equilibrium. The above holds when there are no flows that keep the system off equilibrium. In the more general case when there are such flows, the system may approach a stationary state. Then the second equation (2.60) should be replaced by

$$\tilde{\alpha}_j(\mathbf{p}) = (2\pi)^{-3/2} \int \mathrm{d}^3\mathbf{x}\; \mathrm{e}^{i\,\mathbf{p}\cdot\mathbf{x}}\;[\alpha_j(\mathbf{x},0) - \alpha_j^{(0)}\,(\mathbf{x})], \tag{2.61}$$

where $\alpha_j^{(0)}(\mathbf{x})$ is a stationary solution, satisfying the equation

$$\nabla \cdot (\mathbf{L}_{ik} \cdot (G_{kj} \nabla \alpha_j^{(0)}(\mathbf{x}))) = 0, \tag{2.62}$$

with the proper boundary conditions. The entropy production is (see (2.55), (2.56), (2.57), and (2.59))

$$\dot{S}(t) = -\dot{\Sigma}(t) = G_{ij} \int_V d^3\mathbf{x}\, \alpha_i(\mathbf{x},t)\dot{\alpha}_j(\mathbf{x},t) =$$

$$= -G_{ij} L_{j\gamma k\beta} G_{kl} \int_V d^3\mathbf{x}\, \alpha_i \nabla_\gamma \nabla_\beta \alpha_l =$$

$$= G_{ij} L_{j\gamma k\beta} G_{kl} \int_V d^3\mathbf{x} \frac{\partial \alpha_i}{\partial \mathrm{x}_\gamma} \frac{\partial \alpha_l}{\partial \mathrm{x}_\beta} =$$

$$= \int_V d^3\mathbf{x}\, \mathbf{J}_i \cdot \mathbf{X}_i = \int_V d^3\mathbf{x}\, \sigma(\mathbf{x},t). \tag{2.63}$$

The entropy production density

$$\sigma(\mathbf{x},t) = \mathbf{X}_i(\mathbf{x},t) \cdot \mathbf{L}_{ij} \cdot \mathbf{X}_j(\mathbf{x},t) \geq 0 \tag{2.64}$$

is a near-equilibrium version of (2.45). The time derivative of the entropy production is

$$\ddot{S}(t) = 2\int_V d^3\mathbf{x}\, \mathbf{X}_j \cdot (\mathbf{L}_{jk}\, \dot{\mathbf{X}}_k) = 2G_{ij} L_{j\gamma k\beta} G_{kl} \int_V d^3\mathbf{x} \frac{\partial \alpha_i}{\partial \mathrm{x}_\gamma} \frac{\partial \dot{\alpha}_l}{\partial \mathrm{x}_\beta} =$$

$$= -2G_{ij} L_{j\gamma k\beta} G_{kl} \int_V d^3\mathbf{x} \frac{\partial^2 \alpha_i}{\partial \mathrm{x}_\gamma \partial \mathrm{x}_\beta} \dot{\alpha}_l = -2\int_V d^3\mathbf{x}\, \dot{\alpha}_k G_{kl} \dot{\alpha}_l \leq 0. \tag{2.65}$$

When the system is in a stationary flow which can be described by stationary inhomogeneous boundary conditions then the entropy production decreases towards its minimum. Then the system is described by a solution of (2.63) compatible with the boundary conditions.

2.2 Exergy in Various Systems

A treatment of exergy, which is such a fundamental quantity in the theory of conversions of matter and energy, is usually lacking in thermodynamics text-books. Therefore, we shall here describe the statistical mechanics of a few systems which are of great importance for the study of energy and materials. The discussion is quite conventional (see, e.g., Reif [1965]), apart from our ambition to focus the interest on exergy. We have chosen the following systems: a mixture of ideal gases, a set of models for non-ideal gases, electromagnetic radiation, and an electron gas. The result for the ideal gas mixture will be used in the next section to discuss the emergence of cosmic nuclear exergy in the early Universe.

2.2.1. *Exergy in a mixture of ideal gases*

We consider a gas mixture with M different kinds of non-interacting molecules in a volume V. The average values of the energy and the numbers of molecules are assumed to be U and N_i (i=1,..., M). For free particles the partition function is given by

$$\ln Z = \sum_{i=1}^{M} \ln Z_i = \sum_{i=1}^{M} \ln \sum_{N=0}^{\infty} \frac{e^{-y_i N}}{N!} \sum_{s_1,\ldots,s_N} e^{-\beta(\varepsilon_{s_1}+\ldots+\varepsilon_{s_N})} =$$

$$= \sum_{i=1}^{M} \ln \sum_{N=0}^{\infty} \frac{e^{-y_i N}}{N!} (\zeta_i(\beta))^N = \sum_{i=1}^{M} e^{-y_i} \zeta_i(\beta), \tag{2.66}$$

where $\beta = 1/(kT)$ and $y_i = -\mu_i/(kT)$ are Lagrange multipliers and where

$$\zeta_i(\beta) = \sum_{s} e^{-\beta \varepsilon_s^{(i)}} \tag{2.67}$$

is the partition function for a molecule of kind i. The sum is over molecular states s. The factor $(N!)^{-1}$ enters into (2.66) to compensate for the fact that each state for the molecules of the kind i has been counted $N!$ times since molecules of the same kind are identical particles.

Considering a phase-space cell of volume h^3 as one state, we can write (2.67) as an integral,

$$\zeta_i(\beta) = \frac{V}{h^3} \int d^3\mathbf{p}\, e^{-\beta \mathbf{p}^2/(2m_i)} \sum_s{}' e^{-\beta \varepsilon_s^{(i)}} = V\left(\frac{2\pi m_i}{h^2\beta}\right)^{3/2} \sum_s{}' e^{-\beta \varepsilon_s^{(i)}} . \qquad (2.68)$$

Here, the prime indicates that the sum is over the internal states. If we restrict ourselves to low enough temperatures so that the molecules are not excited, then only the ground state contributes, and

$$\zeta_i = V\left(\frac{2\pi m_i kT}{h^2}\right)^{3/2} \sigma_i , \qquad (2.69)$$

where σ_i is a degeneracy factor for the ground state.

The mean value condition (1.36) with (1.34) applied to (2.66) gives for the equilibrium state the concentrations

$$c_i = \frac{N_i}{V} = -\frac{1}{V}\frac{\partial \ln Z}{\partial y_i} = \frac{1}{V}\, e^{-y_i}\zeta_i . \qquad (2.70)$$

This then fixes y_i and hence μ_i,

$$\mu_i = -kT\, y_i = kT \ln \frac{N_i}{\zeta_i} = kT\left(\ln c_i - \frac{3}{2}\ln\frac{T}{T_o} - \chi_i\right), \qquad (2.71)$$

where T_o is a reference temperature and

$$\chi_i = \frac{3}{2}\ln\left[\frac{1}{h^2}\,(2\pi\, m_i\, k\, T_o\, \sigma_i^{2/3})\right]. \qquad (2.72)$$

Using (1.100), (2.2), and (2.66) we find the pressure

$$p = \frac{kT}{V} \ln Z = k\,T \sum_{i=1}^{M} c_i = \sum_{i=1}^{M} p_i\,, \tag{2.73}$$

where the partial pressures are

$$p_i = k\,T\,c_i\,. \tag{2.74}$$

The relation between energy density and temperature (or pressure) is now obtained from (2.66), (2.69), and (2.73)

$$\frac{U}{V} = -\frac{1}{V}\frac{\partial \ln Z(\beta, y_i)}{\partial \beta} = \frac{1}{V}\sum_{i=1}^{M} \mathrm{e}^{-y_i}\,\zeta_i\left(-\frac{\partial \ln \zeta_i}{\partial \beta}\right) =$$

$$= \sum_{i=1}^{M} c_i \frac{3}{2} k\,T = \frac{3}{2} p\,. \tag{2.75}$$

The entropy is (see (2.5))

$$S = \frac{p\,V}{T} + \frac{U}{T} - \sum_{i=1}^{M} N_i \frac{\mu_i}{T} = k \sum_{i=1}^{M} N_i \left(\frac{3}{2}\ln\frac{T}{T_o} - \ln c_i + \frac{5}{2} + \chi_i\right). \tag{2.76}$$

The equilibrium pressure and chemical potentials are according to (2.71) and (2.74)

$$\begin{cases} p_o = k\,T_o \sum_{i=1}^{M} c_{io}\,, \\ \mu_{io} = k\,T_o\,(\ln c_{io} - \chi_i). \end{cases} \tag{2.77}$$

The exergy (cf. (2.7)) is then

$$E = U + p_o V - T_o S - \sum_{i=1}^{M} \mu_{io} N_i =$$

$$= kT_o \sum_{i=1}^{M} N_i \left[\frac{3}{2} g\left(\frac{T}{T_o}\right) + g\left(\frac{c_{io}}{c_i}\right) \right], \tag{2.78}$$

where $g(x)$ is the function defined in (1.20).

We now assume a fixed number of molecules, N, in the volume V at fixed temperature $T=T_o$. In accordance with this the chemical reactions must conserve the total number of molecules. Then (2.78) reduces to

$$E = k\,T_o V \sum_{i=1}^{M} c_i \ln \frac{c_i}{c_{io}} = k\,T_o\,N\,K\left[\frac{c_o V}{N}; \frac{cV}{N}\right]. \tag{2.79}$$

This equation will be our starting point in Sections 3.2 and 3.3 for analyzing the exergy and information content of chemical systems.

2.2.2 *Exergy in non-ideal gases*

Although the ideal gas approximation is valid and therefore useful in many cases, it is certainly not always possible to use. The most obvious difficulty with the ideal gas model is the absence of a gas-liquid transition. Obviously, the approximation will fail in systems with phase changes.

With few exceptions, the equation of state for a one-component gas can be written as a virial expansion

$$\frac{p}{kT} = \frac{N}{V} + B_2(T)\left(\frac{N}{V}\right)^2 + B_3(T)\left(\frac{N}{V}\right)^3 + \dots \,. \tag{2.80}$$

The coefficients are in general different for different gases. They can be formally derived from knowledge of the intermolecular forces, which, in turn, in principle are given by quantum mechanics. For a sufficiently dilute gas we retain only the first term, i.e., the ideal gas expression.

The technical literature abounds with examples of the use of the exergy concept in systems containing gaseous substances which cannot be treated as ideal gases, and whose properties cannot be derived from a microscopic quantum-mechanical theory with sufficient accuracy. This problem is dealt with by using equation-of-state-expressions where some parameters are experimentally determined. Usually the equations of state are based on the virial expansion (2.86), using heuristically constructed coefficients. For completeness, we shall derive the expression for the exergy in a one-component non-ideal gas, whose equation of state does display a gas-liquid transition.

It is well known that only two things are required for a derivation of any thermodynamic quantity for a pure substance, namely

— one heat capacity, either C_p or C_V, as a function of temperature along any one isobar or isochore, respectively;

— the equation of state.

It is then possible to calculate the entropy and thereby all other thermodynamic functions. We will here exploit this fact to derive a general expression for the exergy of a one-component gas, based on (2.7) in the form

$$E = U - U_{eq} + p_o(V - V_{eq}) - T_o(S - S_{eq}). \qquad (2.81)$$

We assume that the equations of state are written in the form $p=f(V,T)$, where the function f contains parameters that are either experimentally determined through curve-fitting with data or calculated through statistical mechanics. The Maxwell relations can be used to derive the general expressions

$$S(T,V) - S(T_o,V_{eq}) = \int_{T_0}^{T} d\tau \left[\frac{C_V(\tau,V)}{\tau} \right] + \int_{V_{eq}}^{V} d\omega \left(\frac{\partial p}{\partial T} \right)_V (T_o,\omega) , \qquad (2.82)$$

and

$$U(T,V) - U(T_0,V_{eq}) =$$
$$= \int_{T_0}^{T} d\tau C_V(\tau,V) + \int_{V_{eq}}^{V} d\omega \left\{ T_0 \left(\frac{\partial p}{\partial T}\right)_V (T_0,\omega) - p(T_0,\omega) \right\}. \tag{2.83}$$

The remaining term in (2.81) is given directly by the equation of state. Since

$$C_V(T,V) = C_V(T,V_1) + \int_{V_1}^{V} d\omega \left(\frac{\partial C_V(T,\omega)}{\partial \omega}\right)_T , \tag{2.84}$$

where V_1 is the volume on the isochore, and generally

$$\left(\frac{\partial C_V}{\partial V}\right)_T = T\left(\frac{\partial^2 p}{\partial T^2}\right)_V , \tag{2.85}$$

the desired general expression for the exergy is (after simplification)

$$E(T,V) = \int_{T_0}^{T} d\tau \left\{ \left(1 - \frac{T_0}{\tau}\right) \left[C_V(\tau,V_1) + \int_{V_1}^{V} d\omega\, \tau \left(\frac{\partial^2 p(\tau,\omega)}{\partial \tau^2}\right)_V \right] \right\} +$$
$$+ p_0(V - V_{eq}) - \int_{V_{eq}}^{V} d\omega\, p(T_0,\omega) . \tag{2.86}$$

A statistical-mechanical derivation of an equation of state is based upon a microscopic description of interaction potentials etc., which gives the partition function, Z, for the system. Once a partition function has been derived, the equation of state can readily be obtained (cf. (2.78)) by the general relation

$$p = \frac{1}{\beta} \frac{\partial \ln Z}{\partial V} . \tag{2.87}$$

To demonstrate the procedure, we will carry out the calculation for the potential

$$\phi(r) = \begin{cases} \infty & \text{for } r < r^* ; \\ -\phi^* \left(\dfrac{r^*}{r}\right)^6 & \text{for } r > r^* , \end{cases} \tag{2.88}$$

where r is the distance from the center of the weakly attracting hard sphere, and where ϕ^* and r^* are parameters representing the maximum depth of the potential well and the radius of the sphere.

The partition function for a gas of interacting molecules is given by

$$Z = \frac{1}{h^{3N} N!} \left(\frac{2\pi m}{\beta}\right)^{\frac{3N}{2}} \int d^3\mathbf{x}_1 \ldots \int d^3\mathbf{x}_N \, e^{-\beta \Phi(\mathbf{x}_1,\ldots,\mathbf{x}_N)} , \tag{2.89}$$

where Φ is the potential energy of interaction between the molecules. Here, we assume that

$$\Phi(\mathbf{x}_1,\ldots,\mathbf{x}_N) = \frac{1}{2} \sum_{j \neq k} \phi(r_{jk}); \qquad r_{jk} = |\mathbf{x}_j - \mathbf{x}_k| . \tag{2.90}$$

We will denote the integral in (2.89) by $Z_P(\beta)$ in the following. We assume that the average available volume per molecule is so large compared to the volume of the hard-sphere core that the motions of any pair of atoms is effectively uncorrelated with the motion of all other atoms. The average potential energy between an arbitrary pair of molecules j and k is then given by

$$\phi(\beta) = \langle\phi(r_{jk})\rangle = -\frac{\partial}{\partial\beta} \ln \int d^3\mathbf{x}_{jk} \, e^{-\beta\,\phi(r_{jk})} , \tag{2.91}$$

where the integration is over all possible values of $\mathbf{r}_{jk}$, i.e., over the whole volume V. Dropping indices in (2.91), and introducing

$$I(\beta) = \int d^3\mathbf{x}(e^{-\beta\phi(r)} - 1) = 4\pi \int d^3r(e^{-\beta\phi(r)} - 1)r^2 , \qquad (2.92)$$

we find

$$\phi(\beta) = -\frac{\partial}{\partial\beta}\left(\ln V + \ln\left(1 + \frac{I}{V}\right)\right) = -\frac{\partial}{\partial\beta}\left(\frac{I}{V} + \ldots\right) \approx -\frac{1}{V}\frac{\partial I}{\partial\beta} . \qquad (2.93)$$

Furthermore,

$$-\frac{\partial}{\partial\beta} \ln Z_P(\beta) = <\Phi(\mathbf{x}_1,\ldots,\mathbf{x}_N)> = \frac{1}{2} N(N-1)\phi(\beta) \approx \frac{1}{2} N^2\phi(\beta). \qquad (2.94)$$

Thus,

$$\ln Z_P(\beta) = \ln Z_P(0) - \frac{1}{2}N^2 \int_0^\beta d\beta' \ \phi(\beta') = N \ln V + \frac{1}{2}\frac{N^2}{V} I(\beta) . \qquad (2.95)$$

With the potential (2.88) we get

$$I(\beta) = -4\pi \int_0^{r^*} dr\, r^2 + 4\pi \int_{r^*}^{\infty} dr\, (e^{\beta\phi^*(r^*/r)^6} - 1)\, r^2 =$$

$$= -\frac{4\pi}{3} r^{*3} + 4\pi \int_{r^*}^{\infty} dr \left[\beta\phi^*\left(\frac{r^*}{r}\right)^6 + \frac{1}{2}\left(\beta\phi^*\left(\frac{r^*}{r}\right)^6\right)^2 + \ldots\right] r^2 \approx$$

$$\approx \frac{4\pi}{3} r^{*3}(\beta\phi^* - 1) , \qquad (2.96)$$

where we have assumed that the thermal energy kT is large in comparison with ϕ^*. Insertion into (2.87) gives the familiar van der Waals equation

$$p(T,V) = \frac{RT}{\frac{V}{n} - b} - \frac{an^2}{V^2} , \qquad (2.97)$$

where the values of the parameters a and b are connected to the intermolecular potential (2.88) by

$$b = N_A \frac{2\pi r^{*3}}{3}\ ; \qquad a = bN_A\phi^*\ . \tag{2.98}$$

(N_A is Avogadro's number, $R=kN_A$ is the gas constant.)

The total mean energy of the van der Waals' gas is

$$U = -\frac{\partial}{\partial\beta}\ln Z = \frac{3}{2}NkT - \frac{N^2}{VN_A^2}a\ , \tag{2.99}$$

and consequently

$$C_V = \frac{3}{2}Nk\ . \tag{2.100}$$

Insertion in (2.86) of these expressions yields

$$E = \frac{3}{2}nRT_o g\left(\frac{T}{T_o}\right) + nRT_o g\left(\frac{V-nb}{V_{eq}-nb}\right) + a\left(\frac{n}{V_{eq}}\right)^2\left(\frac{V_{eq}^2 - V^2}{V}\right), \tag{2.101}$$

where we again have used the g-function (1.20). Comparison with the ideal-gas expression (2.78) shows that the first term in (2.101) is identical, the second is similar, the volume of the hard spheres has been subtracted, whereas the third term is new. It arises from the attraction between the molecules, and adds to the exergy when $V<V_{eq}$.

Empirical equations of state

The empirical equations of state that are used can often be recognized as truncated and modified virial expansions [Reid et al. 1977]. Many of these empirical expressions owe part of their usefulness to being accurate at representing the properties of gaseous mixtures. In particular, expressions with few, say two, parameters are useful, since for them it is possible to construct mixing rules from microscopic theory. One of the most widely used expressions, suggested by Redlich and Kwong [1949], can be written as

$$p = \frac{nRT}{V - bn} - \frac{an^2}{V(V + bn)\sqrt{T}}, \tag{2.102}$$

where a and b are the only parameters (see also Soave [1972]). The resulting exergy expression is

$$E = \int_{T_o}^{T} d\tau \left[\left(1 - \frac{T_o}{\tau}\right) C_V(\tau, V_1) \right] + \frac{an}{2b\sqrt{T}} \left(3 - 2\left(\frac{T}{T_o}\right)^{\frac{1}{2}} - \frac{T_o}{T}\right) \ln\left(\frac{V_1(V+bn)}{V(V_1+bn)}\right)$$

$$+ nRT_o\, g\left(\frac{V - bn}{V_{eq} - bn}\right) + \frac{an}{b\sqrt{T_o}} \left[g\left(\frac{V + bn}{V_{eq} + bn}\right) - g\left(\frac{V}{V_{eq}}\right) \right]. \tag{2.103}$$

We may again recognize the ideal-gas terms. In this case the effects of the non-ideal terms are not easily described, but it is clear that they may be positive or negative. Our examples thus show that using the ideal-gas approximation may lead to under-estimating the exergy as well as to over-estimating it.

2.2.3 *Exergy in electromagnetic radiation*

We shall now consider a gas of free photons within a certain volume V. There is no restriction on the number of photons; hence there is no chemical potential. The logarithm of the partition function for this system can be written as a spectral integral,

$$\alpha(\beta) = \ln Z(\beta) = V \int_0^\infty d\varepsilon\, \rho(\varepsilon) \ln z(\beta\varepsilon). \tag{2.104}$$

Here $z(\beta\varepsilon)$ is the partition function for photons in a state of energy ε at temperature $1/(k\beta)$,

$$z(y) = \sum_{n=0}^{\infty} \mathrm{e}^{-ny} = \frac{1}{1-\mathrm{e}^{-y}}\,; \quad y = \beta\varepsilon\,, \tag{2.105}$$

and

$$\rho(\varepsilon) = \frac{8\,\pi}{c^3\,h^3}\,\varepsilon^2 \tag{2.106}$$

(with c = the velocity of light and h = Planck's constant) is the number of photon states per unit energy interval and unit volume. The probability distribution over photon number for a given photon energy is then at equilibrium

$$p_n = \frac{1}{z(y)}\,\mathrm{e}^{-ny} = \mathrm{e}^{-ny} - \mathrm{e}^{-(n+1)y}\,, \tag{2.107}$$

which gives the mean occupation number

$$< n > = \sum_{n=0}^{\infty} n\,p_n = -\frac{z'(y)}{z(y)} = z(y) - 1 = \frac{1}{\mathrm{e}^{y} - 1} = \frac{1}{\mathrm{e}^{\beta\varepsilon} - 1}\,. \tag{2.108}$$

Inserting now (2.105) and (2.106) in (2.104) we find for the logarithm of the partition function

$$\alpha(\beta) = \frac{8\pi V}{c^3\,h^3}\int_0^{\infty} \mathrm{d}\varepsilon\,\varepsilon^2 \ln\frac{1}{1-\mathrm{e}^{-\beta\varepsilon}} = \frac{8\,\pi\,V}{(ch\beta)^3}\int_0^{\infty} \mathrm{d}y\,y^2 \ln\frac{1}{1-\mathrm{e}^{-y}} =$$

$$= \frac{8\pi V}{3(ch\beta)^3}\int_0^{\infty} \mathrm{d}y\,\frac{y^3}{\mathrm{e}^{y} - 1} = \frac{8\pi V}{3(ch\beta)^3}\cdot\frac{\pi^4}{15} = \frac{\pi^2}{45(c\hbar)^3}\cdot\frac{V}{\beta^3}\,; \quad \left(\hbar = \frac{h}{2\pi}\right). \tag{2.109}$$

The energy density of the photon gas for thermal equilibrium at temperature $T = 1/(k\beta)$ (black-body radiation) is

$$u(T) = \frac{1}{V}\,U(\beta) = -\frac{1}{V}\,\frac{\partial\,\alpha(\beta)}{\partial\beta} = 3\,\frac{k\,T}{V}\,\alpha(\beta) = \frac{\pi^2\,k^4}{15(c\hbar\,)^3}\,T^{\,4}. \tag{2.110}$$

This presupposes interaction with absorbtive/emissive (black) matter acting as a heat bath at temperature T.

The entropy of black-body radiation is (see (1.95))

$$\frac{S}{V} = \frac{k}{V}\,[\alpha(\beta) + \beta\,U(\beta)] = \frac{4}{3}\,\frac{u(T)}{T}\,, \tag{2.111}$$

and the radiation pressure is (see (2.2))

$$p = \frac{k\,T}{V}\,\alpha(\beta) = \frac{1}{3}\,u(T). \tag{2.112}$$

The mean number of photons is obtained from (2.108) summed over all states using (2.106),

$$\begin{aligned}\frac{N_\gamma}{V} &= \int_0^\infty \mathrm{d}\varepsilon\,\frac{\rho(\varepsilon)}{\mathrm{e}^{\beta\varepsilon}-1} = \frac{8\,\pi}{(ch\beta)^3}\int_0^\infty \mathrm{d}y\,\frac{y^{\,2}}{\mathrm{e}^y - 1} = \frac{16\pi\,\zeta(3)}{(ch\beta)^3} = \\ &= \frac{2\,\zeta(3)}{\pi^2}\left(\frac{k\,T}{c\hbar}\right)^3,\end{aligned} \tag{2.113}$$

where $\zeta(t)$ is the Riemann zeta-function. Numerically, $\zeta(3) \approx 1.202$.

With a reference state consisting of black-body radiation of the same volume V at a different temperature T_0, the exergy density of our photon gas is

$$\begin{aligned}\frac{E}{V} &= u(T) - u(T_0) - T_0\,\frac{4}{3}\left(\frac{u(T)}{T} - \frac{u(T_0)}{T}\right) = \\ &= u(T)\left(1 - \frac{4}{3}\,\frac{T_0}{T} + \frac{1}{3}\left(\frac{T_0}{T}\right)^4\right)\end{aligned} \tag{2.114}$$

To generalize the discussion to arbitrary incoherent radiation, we note that according to (2.108)

$$y = \ln(<n>^{-1} + 1), \tag{2.115}$$

and the probability (2.107) may be written (cf. (1.56))

$$p_n = \frac{1}{<n>+1}\left(\frac{1}{<n>}+1\right)^{-n} = \frac{<n>^n}{(<n>+1)^{n+1}}. \tag{2.116}$$

What characterizes incoherent electromagnetic radiation is that for each mode, the probability distribution is of the form (2.116). We let **k** denote the wave vector and σ polarization. The photon energy is

$$\varepsilon = c\hbar\,|\mathbf{k}|. \tag{2.117}$$

We let $n(k,\sigma)$ be the mean number of photons per state for the mode $(\mathbf{k},\sigma)$, and we introduce (cf. (2.108))

$$\beta(\mathbf{k},\sigma) = \frac{1}{c\hbar|\mathbf{k}|}\ln\left(\frac{1}{n(\mathbf{k},\sigma)}+1\right), \tag{2.118}$$

thus $[k\beta(\mathbf{k},\sigma)]^{-1}$ is an equivalent temperature for the particular mode $(\mathbf{k},\sigma)$. We can now determine the exergy as a sum over exergies of the different degrees of freedom. With the same reference state as above ($\beta_o=1/(kT_o)$) we obtain from (1.114) and (2.107),

$$E = \frac{V}{(2\pi)^3}\cdot\frac{1}{\beta_o}\sum_\sigma\int d^3\mathbf{k}\sum_{n=0}^{\infty} e^{-\beta n\varepsilon}(1-e^{-\beta\varepsilon})\ln\frac{e^{-\beta n\varepsilon}(1-e^{-\beta\varepsilon})}{e^{-\beta_o n\varepsilon}(1-e^{-\beta_o\varepsilon})}, \tag{2.119}$$

which gives the exergy density

$$\frac{E}{V} = (2\pi)^{-3}\sum_\sigma\int d^3\mathbf{k}\; n(\mathbf{k},\sigma)\, c\hbar\,|\mathbf{k}|\; q(\mathbf{k},\sigma). \tag{2.120}$$

Here $q(\mathbf{k},\sigma)$ is an energy quality function [Karlsson 1982]

$$q(\mathbf{k},\sigma) = 1 - \frac{\beta}{\beta_o} + \frac{e^{\beta\varepsilon} - 1}{\beta_o\varepsilon} \ln \frac{1 - e^{-\beta\varepsilon}}{1 - e^{-\beta_o\varepsilon}} . \tag{2.121}$$

Finally, we shall consider a more general kind of radiation field with a coherent field described the amplitude $a(\mathbf{k},\sigma)$ superposed on field described by the photon density $b(\mathbf{k},\sigma)$. Then the exergy is still of the form (2.120), where the photon density is

$$n(\mathbf{k},\sigma) = b(\mathbf{k},\sigma) + |a(\mathbf{k},\sigma)|^2. \tag{2.122}$$

The energy quality now is

$$q = 1 - \frac{\beta b}{\beta_o n} + \frac{1}{\beta_o n\varepsilon} \ln \frac{1 - e^{-\beta\varepsilon}}{1 - e^{-\beta_o\varepsilon}} ; \quad \beta = \frac{1}{\varepsilon} \ln \frac{b + 1}{b} . \tag{2.123}$$

To show this, one notices that for each mode the coherent radiation, which has zero entropy, contributes a separate term to the exergy density. Thus,

$$\frac{E}{V} = (2\pi)^{-3} \sum_{\sigma} \int d^3\mathbf{k} \left[|a|^2 \varepsilon + \left(1 - \frac{\beta}{\beta_o}\right) b\varepsilon + \frac{1}{\beta_o} \ln \frac{1 - e^{-\beta\varepsilon}}{1 - e^{-\beta_o\varepsilon}} \right]. \tag{2.124}$$

This is a quite general expression for the exergy of radiation.

2.2.4 *The exergy of an electron gas*

We shall consider a gas of free electrons in a fixed volume V, and assume that their kinematical and dynamical situation is characterized by a spectral density $\rho(\varepsilon)$, which will not be specified until later. The logarithm of the partition function is (cf. (2.4))

$$\alpha(\beta,\mu) = \ln Z(\beta,\mu) = V \int_0^\infty \mathrm{d}\varepsilon\, \rho(\varepsilon) \ln z(\beta(\varepsilon - \mu)) =$$

$$= V \frac{1}{\beta} \int_{-\beta\mu}^{\infty} \mathrm{d}y\, \rho\left(\mu + \frac{y}{\beta}\right) \ln z(y), \tag{2.125}$$

where μ is the chemical potential and where z is the partition function for a specific energy (given β and μ),

$$z(y) = 1 + \mathrm{e}^{-y}; \quad y = \beta\,(\varepsilon - \mu). \tag{2.126}$$

The probability distribution for no electron or one electron corresponding to this is

$$\begin{cases} p_0(y) = \dfrac{1}{1 + \mathrm{e}^{-y}}\,; \\[2ex] p_1(y) = \dfrac{1}{1 + \mathrm{e}^{y}}\,. \end{cases} \tag{2.127}$$

For a set of N free electrons distributed over n ($\geq N$) states with the same energy (and hence the same y), we thus obtain again the Bernoulli distribution (1.65) with $c=N/n$. In general we have to take into account the full energy spectrum. The average density of electrons is then

$$\frac{N}{V} = \int_0^\infty \mathrm{d}\varepsilon\, \frac{\rho(\varepsilon)}{1 + \mathrm{e}^{\beta\,(\varepsilon - \mu)}}\,. \tag{2.128}$$

For temperatures approaching zero ($\beta \to \infty$), $p_1(\varepsilon) = 1$ for $\varepsilon < \mu_F$, the *Fermi energy*, and $p_1(\varepsilon)$ vanishes for $\varepsilon < \mu_F$. Here μ_F is determined by

$$\frac{N}{V} = \psi(\mu_F), \tag{2.129}$$

where ψ is the integrated spectral function

$$\psi(\varepsilon) = \int_0^{\varepsilon} \mathrm{d}\varepsilon' \, \rho(\varepsilon'). \tag{2.130}$$

We shall also use the twice integrated spectral function,

$$\chi(\varepsilon) = \int_0^{\varepsilon} \mathrm{d}\varepsilon' \, \psi(\varepsilon'). \tag{2.131}$$

Integrating by parts we can then rewrite (2.125) on the form

$$\alpha(\beta,\mu) = V \int_{-\beta\mu}^{\infty} \mathrm{d}y \, \frac{\psi(\mu + y/\beta)}{1+\mathrm{e}^{y}} = V\beta \int_{-\beta\mu}^{\infty} \mathrm{d}y \, \frac{\chi(\mu + y/\beta)}{(1+\mathrm{e}^{y})\,(1+\mathrm{e}^{-y})} \,. \tag{2.132}$$

In most cases of interest μ will be of the same order of magnitude as μ_{F}, which in turn will be large compared to β^{-1}. The function $\mathrm{e}^{y}(1+\mathrm{e}^{y})^{-2}$ in (8) differs considerably from zero only in interval of a few units' width around y=0. The function χ is relatively slowly varying. To extend the algebraic definition of χ to negative values and to extend the lower limit of integration to -∞ introduces an error of relative order $\mathrm{e}^{-\beta\mu_{\mathrm{F}}}$, which is negligible. It is therefore customary to approximate (2.132) by

$$\alpha(\beta,\mu) = V\beta \int_{-\infty}^{\infty} \mathrm{d}y \, \frac{\chi(\mu + y/\beta)}{(1+\mathrm{e}^{y})\,(1+\mathrm{e}^{-y})} \,. \tag{2.133}$$

The electron density (2.129) can also be written (for finite β)

$$\psi(\mu_{\mathrm{F}}) = \frac{N}{V} = \frac{1}{V\beta}\frac{\partial}{\partial\mu}\,\alpha(\beta,\mu) = \int_{-\infty}^{\infty} \mathrm{d}y \, \frac{\psi(\mu + y/\beta)}{(1+\mathrm{e}^{y})\,(1+\mathrm{e}^{-y})} \,. \tag{2.134}$$

This gives a relation between μ, β, and N/V (or μ_F). The energy density is

$$\frac{U}{V} = -\frac{1}{V}\left(\frac{\partial\alpha}{\partial\beta} - \frac{\mu}{\beta}\frac{\partial\alpha}{\partial\mu}\right) = \int_{-\infty}^{\infty} \mathrm{d}y \, \frac{\left(\mu + \frac{y}{\beta}\right)\psi\left(\mu + \frac{y}{\beta}\right) - \chi\left(\mu + \frac{y}{\beta}\right)}{(1+\mathrm{e}^{y})\,(1+\mathrm{e}^{-y})} . \tag{2.135}$$

The pressure of the electron gas is (see the second equation (2.2))

$$p(\beta,\mu) = \frac{kT}{V}\,\alpha(\beta,\mu) = \int_{-\infty}^{\infty} \mathrm{d}y \, \frac{\chi(\mu + y/\beta)}{(1+\mathrm{e}^{y})\,(1+\mathrm{e}^{-y})} . \tag{2.136}$$

The entropy density is then obtained from the Gibbs relation

$$\frac{S}{V} = \frac{1}{VT}\,(U + pV - \mu N) = k\beta\left(\frac{U}{V} + p - \mu\frac{N}{V}\right) =$$

$$= k \int_{-\infty}^{\infty} \mathrm{d}y \, \frac{y\psi(\mu + y/\beta)}{(1+\mathrm{e}^{y})\,(1+\mathrm{e}^{-y})} . \tag{2.137}$$

With a reference state characterized by β_o and μ_o and the corresponding density N_o/V and Fermi energy μ_{Fo},

$$\psi(\mu_{Fo}) = \frac{N_o}{V} = \int_{-\infty}^{\infty} \mathrm{d}y \, \frac{\psi(\mu_o + y/\beta_o)}{(1+\mathrm{e}^{y})\,(1+\mathrm{e}^{-y})} , \tag{2.138}$$

the exergy density is (see (2.7))

$$\frac{E}{V} = \frac{U}{V} + p_o - T_o\frac{S}{V} - \mu_o\frac{N}{V} =$$

$$= \int_{-\infty}^{\infty} \frac{\mathrm{d}y}{(1+\mathrm{e}^{y})(1+\mathrm{e}^{-y})}\left[\left(\mu + \frac{y}{\beta} - \mu_o - \frac{y}{\beta_o}\right)\psi\left(\mu + \frac{y}{\beta}\right)\right.$$

$$\left. - \chi\left(\mu + \frac{y}{\beta}\right) + \chi\left(\mu_o - \frac{y}{\beta_o}\right)\right] . \tag{2.139}$$

For a *free electron gas in two dimensions*, ρ is constant, and

$$\begin{cases} \rho(\varepsilon) = 4\pi\, m\,; \\ \psi(\varepsilon) = 4\pi\, m\, \varepsilon\,; \\ \chi(\varepsilon) = 2\pi\, m\, \varepsilon^2\,. \end{cases} \tag{2.140}$$

The y-integrals have the numerical values,

$$\int_{-\infty}^{\infty} \frac{\mathrm{d}y}{(1+\mathrm{e}^{y})\,(1+\mathrm{e}^{-y})}\,(1, y, y^2) = \left(1, 0, \frac{\pi^2}{3}\right). \tag{2.141}$$

In this case (2.134), (2.135), (2.137), and (2.139) simplify to

$$\begin{cases} \mu = \mu_{\mathrm{F}} = \dfrac{1}{4\pi\, m}\cdot\dfrac{N}{V}\,; \\ \dfrac{U}{V} = 2\pi\, m\left(\mu_{\mathrm{F}}^2 + \dfrac{\pi^2}{3}\,(kT)^2\right); \\ \dfrac{S}{V} = \dfrac{4\,\pi^3}{3}\, m\, k^2\, T\,; \\ \dfrac{E}{V} = \dfrac{2\,\pi^3}{3}\, m\,(k\,(T - T_{\mathrm{o}}))^2. \end{cases} \tag{2.142}$$

In the general case one can expand the spectral functions around μ_{F} (the so-called Sommerfeld expansion) in powers of x/β. Here μ_{F} is determined by (2.134). The expansion should converge rapidly for $\beta\mu \ll 1$. Then we get to second order in kT/μ

$$\begin{cases} \mu = \mu_F - \dfrac{\pi^2}{6} \cdot \dfrac{\rho'(\mu_F)}{\rho(\mu_F)} (kT)^2 ; \\ U = U_F + \dfrac{\pi^2}{6} \rho(\mu_F) (kT)^2 ; \\ S = \dfrac{\pi^2}{6} \rho(\mu_F) k^2 T ; \\ E = \dfrac{\pi^2}{6} \rho(\mu_F) (k (T - T_o))^2 . \end{cases} \tag{2.143}$$

In addition to the conventional relations for an electron gas we have thus obtained an expression for the exergy.

2.3 Cosmic Nuclear Exergy

It is an old theme of cosmic thermodynamics that the second law implies that the Universe is running down, i.e., running out of exergy and approaching thermodynamic equilibrium, 'heat death.' However, in the standard model of the early universe [Weinberg 1972], the Universe had already reached a state of heat death 0.01s after Big Bang—disregarding the gravitons which were already effectively decoupled from the rest. Nucleons, electrons and positrons, neutrinos, and photons were in equilibrium. The problem is then to understand the revival of the Universe. How was exergy created, in particular the nuclear exergy of the Sun and other stars?

The answer lies in the fact that the Universe was not static [Tolman 1934, Davies 1974, Layzer 1975, Landsberg 1978]. It was expanding, thus changing the conditions for equilibrium. We shall see how the nuclear exergy of the Universe started to form after around 10s. The process was essentially completed in 24h—the first day of creation. The nucleon-nucleus gas was during this time non-relativistic and non-degenerate.

We shall derive an expression for the exergy of nucleons and nuclei following Eriksson, Islam, and Skagerstam [1982]. Then we need to know the equilibrium state of the gas of nucleons and nuclei. The era of interest was dominated by radiation, i.e., most of the energy was in the form of radiation. This radiation was then acting as a heat bath which determined the temperature $T = 1/(k\beta)$ of the gas of nuclides.

To describe microstates of the nuclide gas we introduce the vector

$$\boldsymbol{n} = (n_{ik}), \tag{2.144}$$

where n_{ik} is the number of nuclides of type i in the state k. Let

$$\begin{cases} \boldsymbol{A} = (A_{ik})\ ; \quad A_{ik} = A_i\ , \quad \text{all } k; \\ \boldsymbol{u} = (u_{ik}), \end{cases} \tag{2.145}$$

be the corresponding vectors for nucleon number and energy. Then for a given energy U and a given number of nucleons N,

$$\begin{cases} <\boldsymbol{n}\cdot\boldsymbol{u}> = U\ ; \\ <\boldsymbol{n}\cdot\boldsymbol{A}> = N\ . \end{cases} \tag{2.146}$$

We use the fact that at the relevant densities and pressures the nuclides were non-degenerate and non-relativistic. We neglect internal excitations and interactions. The result is then an ideal gas model, similar to the one in Section 2.2.1. The equilibrium state is given by the Gibbs distribution

$$P(\boldsymbol{n}) = \mathrm{e}^{-\alpha - \beta \boldsymbol{n}\cdot\boldsymbol{\varepsilon} + \beta\mu_o \boldsymbol{n}\cdot\boldsymbol{A}}\ , \tag{2.147}$$

where α is the logarithm of the partition function

$$\alpha(\beta, \mu_o) = \ln Z(\beta, \mu_o) = \ln\left(\sum_{\boldsymbol{n}} \frac{e^{\beta\, \boldsymbol{n}\cdot(\mu_o \boldsymbol{A} - \boldsymbol{\varepsilon})}}{\prod_{i,k} n_{ik}!}\right) =$$

$$= \sum_i V\left(\frac{2\pi\, A_i\, m}{h^2\, \beta}\right)^{3/2} g_i\, \mathrm{e}^{\beta A_i (\mu_o + b_i)}\ . \tag{2.148}$$

Here we have used a uniform state density h^3 in phase space; k has been replaced by position **p** in momentum space, and

$$\varepsilon_i(\mathbf{p}) = -A_i\, b_i + \frac{\mathbf{p}^2}{2\, m\, A_i}, \tag{2.149}$$

where m is the nucleon mass, and b_i is the binding energy per nucleon of the nuclide i. Finally g_i in (2.148) is spin degeneracy of nuclide i.

Further, since only the average total number of nucleons has been fixed and not the average numbers of the various nuclides—which means that transitions between various nuclides are considered as possible through nuclear reactions—there is only one chemical potential μ_o appearing in (2.147) and (2.148).

The volume in (2.148) is

$$V = \frac{N}{\theta \cdot \dfrac{N_\gamma}{V}} = N\, \frac{\pi^2}{2\, \zeta(3)\, \theta}\, (\hbar\, c\, \beta)^3, \tag{2.150}$$

where θ is the nucleon to photon ratio, and where we have used (2.113). This inserted in (2.148) gives

$$\alpha(T, \mu_o) = N \cdot \eta \left(\frac{m\, c^2}{k T} \right)^{3/2} \sum_i g_i\, A_i^{3/2}\, e^{A_i (\mu_o + b_i)/k T}; \; \eta = \frac{\sqrt{2\pi}}{8\, \zeta(3)\, \theta}. \tag{2.151}$$

It is a consistency requirement that $N = kT(\partial\alpha/\partial\mu_o)$, hence

$$\eta \left(\frac{m\, c^2}{k T} \right)^{3/2} \sum_i g_i\, A_i^{5/2}\, e^{A_i (\mu_o + b_i)/k T} = 1. \tag{2.152}$$

This fixes μ_o as a function of temperature. The equilibrium probability for a randomly picked nucleon to be in a nuclide A_i is then

$$q_i^{(o)}(T) = \eta \left(\frac{m\, c^2}{k T} \right)^{3/2} g_i\, A_i^{5/2}\, e^{A_i (\mu_o(T) + b_i)/k T}. \tag{2.153}$$

A simple model

The nuclide gas includes only proton, neutron, ^{4}He, and ^{56}Fe (^{4}He because it was actually formed during the era considered, ^{56}Fe because it is the most tightly bound nuclide) and neglect all other nuclides. The numerical values used are mc^2 = 938.8 MeV and η = 0.26×10^9 (based on the estimate θ= 10^{-9}), and those values given in Table 2.1.

Table 2.1. Characteristics for the nuclides used in the model for the nucleon gas in the early universe.

i	$^{A_i}Z_i$	g_i	b_i (MeV)
1	^{1}H	2	0
2	^{1}n	2	−1.30
3	^{4}He	1	6.43
4	^{56}Fe	1	8.10

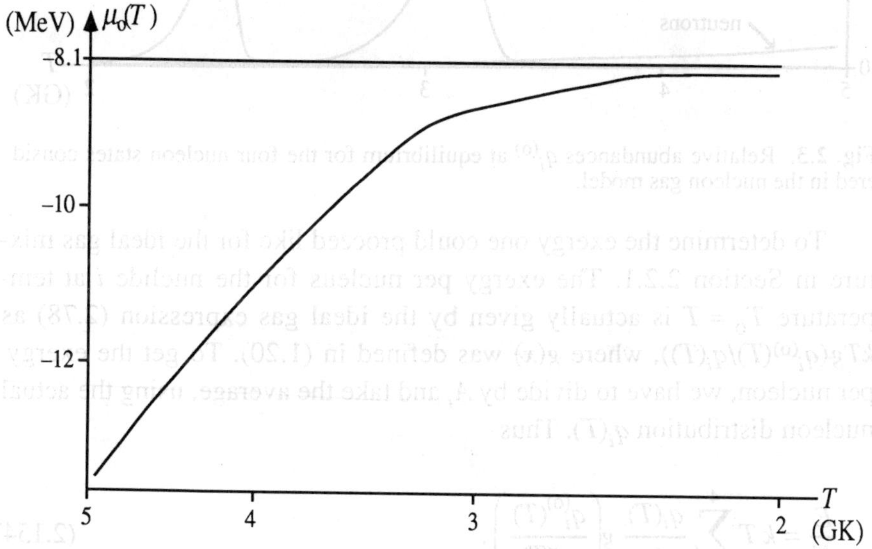

Fig. 2.2. The nucleon equilibrium chemical potential as a function of temperature for a model of the gas of nucleons and nuclei in the early universe.

Solving for $\mu_o(T)$ in (2.152), we find the behaviour shown in Fig. 2.2 for decreasing temperature. The asymptotic value, $-b_4$, is reached for $T \approx 2$ GK. This gives, within the model, the equilibrium nuclide distribution (2.153), plotted in Fig. 2.3. The changes here, first from hydrogen to helium and then to iron, are reflected in changes in $d\mu_o(T)/dT$, which can be seen in Fig. 2.2.

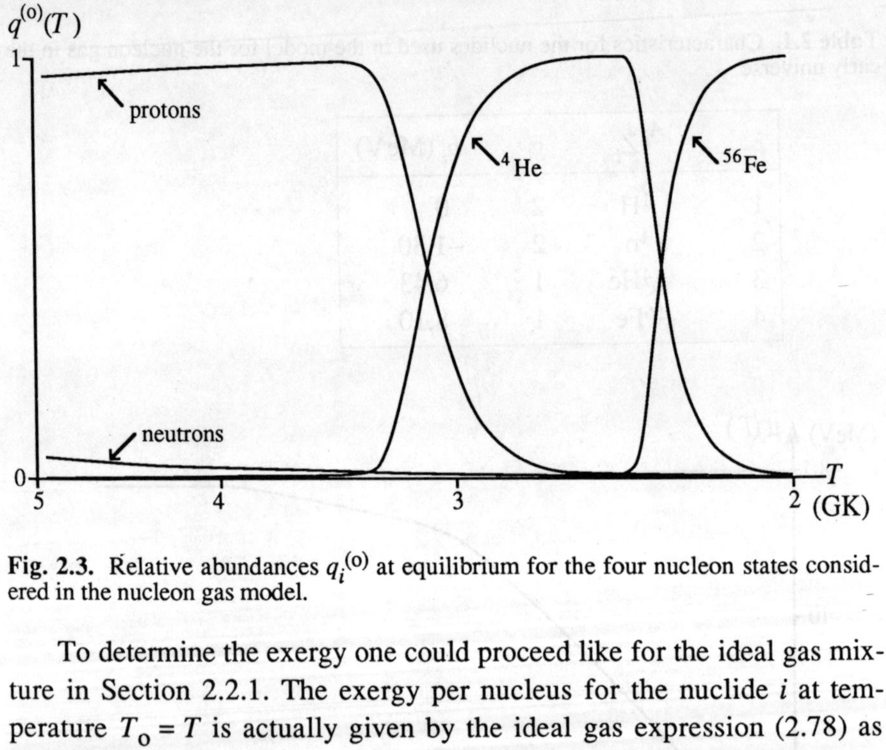

Fig. 2.3. Relative abundances $q_i^{(o)}$ at equilibrium for the four nucleon states considered in the nucleon gas model.

To determine the exergy one could proceed like for the ideal gas mixture in Section 2.2.1. The exergy per nucleus for the nuclide i at temperature $T_o = T$ is actually given by the ideal gas expression (2.78) as $kTg(q_i^{(o)}(T)/q_i(T))$, where $g(x)$ was defined in (1.20). To get the exergy per nucleon, we have to divide by A_i and take the average, using the actual nucleon distribution $q_i(T)$. Thus

$$\frac{E}{N} = kT \sum_{i=1}^{4} \frac{q_i(T)}{A_i}\, g\left(\frac{q_i^{(o)}(T)}{q_i(T)}\right). \tag{2.154}$$

For simplicity, the helium synthesis is modelled as occuring instantaneously, at a time when the temperature was T_0. All neutrons then available are assumed to go into ^{4}He. Then $q_i(T)$ may be expressed in terms of the neutron fraction $\gamma(T)$ as in Table 2.2.

Table 2.2. Nucleon distribution, $q_i(T)$ ($i = 1,\ldots, 4$), over the four nuclides of the model.

i	$T > T_0$	$T < T_0$
1	$1-\gamma(T)$	$1-2\gamma(T)$
2	$\gamma(T)$	0
3	0	$2\gamma(T)$
4	0	0

For neutron fraction $\gamma(T)$ we have used the values of Johansson et al. [1982]. (See also Peebles [1966].) For T_0 we have chosen the value 10^9 K and for $\gamma(T_0)$ the value 0.125. After the helium synthesis at T_0= 0.81 GK, the exergy per nucleon is approximately

$$\frac{E}{N} = 6.49 \text{ MeV} - kT(15.46 + 1.192 \ln \frac{mc^2}{kT}).$$

In a cooling universe the temperature may serve as a time parameter. We have plotted the exergy per nucleon as a function of decreasing T in Fig. 2.4.

As an attractive force, gravitation is inherently unstable against density fluctuations, if the pressure does not force matter to stay homogeneous. Therefore, when pressure dropped, in particular when plasma turned into a gas of atoms around the time 700,000 years, gravitational clumping became possible. Thus gravitational exergy was created. As an example, in a protostar which is to end up as a cold neutron star and a supernova cloud, the nuclear-gravitational exergy per nucleon is around 40 MeV, most of it gravitational. The very heavy elements, such as uranium, thus have a nuclear exergy which is mainly of gravitational origin.

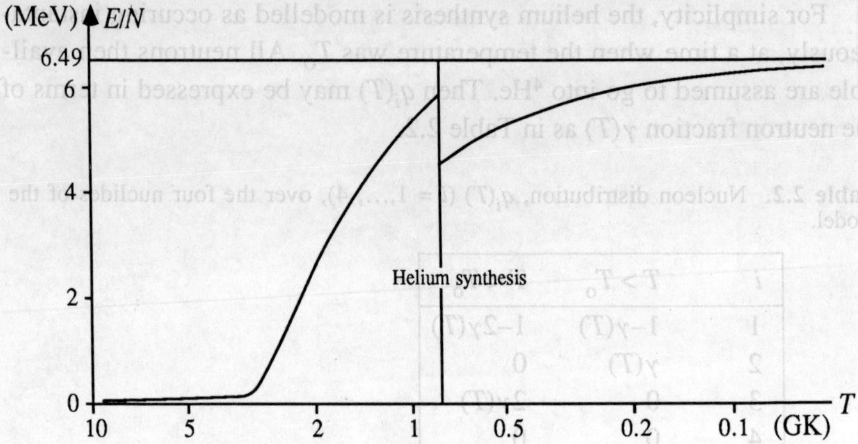

Fig. 2.4. The exergy per nucleon in the early universe. The exergy is being built up to an asymptotic value of 6.49 MeV, apart from a sharp drop at $T=T_o$, where ^{4}He is formed.

Many themes would be interesting to study in this context: black holes, the future fate of the universe [Frautschi 1982], proton decay, heat deaths which took place earlier than that of 0.01s [Van Hove 1983], but we shall leave the theme here.

2.4 Optimized Exergy Extraction

Carnot's problem was to determine the amount of work obtainable from and engine operating in a heat flow between a heat source and a heat sink. This situation can be generalized to an arbitrary number of heat sources (or sinks), assuming the flows to be limited to certain intervals [Grubbström 1980]. Moreover, it can be generalized to other sources than just heat, such as sources of chemical reactants [Eriksson 1984]. These are formalized problems which are quite close to applications in many fields, such as process optimization.

2.4.1 *Mechanical power from a set of heat sources with limited flows*

Let us assume that there are M heat sources (or sinks) with different temperatures $0 < T_1 < T_2 < ... < T_M$. The heat flows q_i may go from the heat sources or into them and are accordingly counted as positive or negative respectively. They are assumed to be bounded within certain limits,

$$\check{q}_i \le q_i \le \hat{q}_i\,; \quad \check{q}_i < \hat{q}_i\,; \quad (i = 1,..,M). \tag{2.155}$$

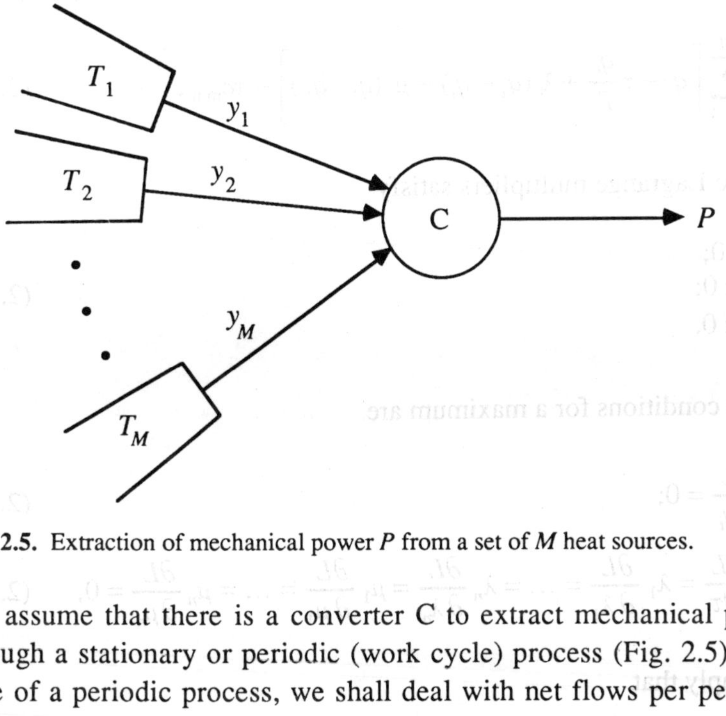

Fig. 2.5. Extraction of mechanical power P from a set of M heat sources.

We assume that there is a converter C to extract mechanical power P through a stationary or periodic (work cycle) process (Fig. 2.5). For the case of a periodic process, we shall deal with net flows per period. We further assume that the rate of entropy production in the converter has a lower limit $\sigma_{\min}$. Then from energy conservation and entropy limitation we have

$$\begin{cases} P = \sum_{i=1}^{M} q_i \,; & (2.156) \\ -\sum_{i=1}^{M} \frac{q_i}{T} \geq \sigma_{\min} \,. & (2.157) \end{cases}$$

The formal solution is to maximize (2.156) under the constraints (2.155) and (2.157), assumed to be compatible. For this we form the Lagrangian

$$L = \sum_{i=1}^{M} \left[q_i - \tau \frac{q_i}{T_i} + \lambda_i \, (\hat{q}_i - q_i) + \mu_i \, (q_i - \breve{q}_i \,) \right] - \tau \sigma_{\min} \,, \tag{2.158}$$

where the Lagrange multipliers satisfy

$$\begin{cases} \tau \geq 0; \\ \lambda_i \geq 0; \\ \mu_i \geq 0. \end{cases} \tag{2.159}$$

Then the conditions for a maximum are

$$\begin{cases} \dfrac{\partial L}{\partial q_i} = 0; & (2.160) \\ \tau \dfrac{\partial L}{\partial \tau} = \lambda_1 \dfrac{\partial L}{\partial \lambda_1} = \ldots = \lambda_n \dfrac{\partial L}{\partial \lambda_n} = \mu_1 \dfrac{\partial L}{\partial \mu_1} = \ldots = \mu_n \dfrac{\partial L}{\partial \mu_n} = 0, & (2.161) \end{cases}$$

which imply that

$$\begin{cases} 1 - \dfrac{\tau}{T_i} - \lambda_i + \mu_i = 0; \\ \tau \left(\sum_{i=1}^{M} \dfrac{q_i}{T_i} + \sigma_{\min} \right) = 0, \end{cases} \tag{2.162}$$

and

$$\begin{cases} \lambda_i (\hat{q}_i - q_i) = 0; & (i = 1,\ldots, M); \\ \mu_i (q_i - \check{q}_i) = 0; & (i = 1,\ldots, M). \end{cases} \tag{2.163}$$

Then, if $\tau > 0$, for each index i one of the following three possibilities is realized:

$$\text{(A)} \begin{cases} T_i = \tau; \\ \lambda_i = 0; \\ \mu_i = 0; \\ \check{q}_i < q_i < \hat{q}_i; \end{cases} \quad \text{or (B)} \begin{cases} T_i < \tau; \\ \lambda_i = 0; \\ \mu_i = \dfrac{\tau}{T_i} - 1; \\ q_i = \check{q}_i; \end{cases} \quad \text{or (C)} \begin{cases} T_i > \tau; \\ \lambda_i = 1 - \dfrac{\tau}{T_i}; \\ \mu_i = 0; \\ q_i = \hat{q}_i. \end{cases} \tag{2.164}$$

The whole problem can be solved only if τ is equated to some T_i, say T_k,

$$\tau = T_k. \tag{2.165}$$

Then for $i = k$, the possibility (A) is realized; for $i < k$, $T_i < T_k = \tau$, and (B) is realized, while for $i > k$, $T_i > \tau$ and (C) is realized. Thus we have

$$\begin{cases} \lambda_k = \mu_k = 0; \\ \check{q}_k \leq q_k \leq \hat{q}_k, \end{cases} \tag{2.166}$$

and

$$\begin{cases} \mu_i = \dfrac{T_k}{T_i} - 1; \\ \lambda_i = 0; \\ q_i = \check{q}_i; \end{cases} \quad (i < k) \qquad \begin{cases} \lambda_i = 1 - \dfrac{T_k}{T_i}; \\ \mu_i = 0; \\ q_i = \hat{q}_i. \end{cases} \quad (i > k) \tag{2.167}$$

Thus, if $\tau > 0$, the heat source k serves as a heat bath for the whole process. The heat sources with $i > k$ are 'boilers' and those with $i < k$ serve as 'coolers.' From (2.165) and (2.162) we conclude that the entropy production is at its minimum,

$$\dot{S} = -\sum_{i=1}^{M} \frac{q_i}{T_i} = \sigma_{\min} . \tag{2.168}$$

The extracted mechanical power may now be written

$$P = \sum_{i=1}^{M} q_i = \sum_{i=k+1}^{M} \hat{q}_i \left(1 - \frac{T_k}{T_i}\right) - \sum_{i=1}^{k-1} \breve{q}_i \left(\frac{T_k}{T_i} - 1\right) - T_k\, \sigma_{\min} . \tag{2.169}$$

We see that the flows are valued according to their Carnot factors and that T_k serves as a reference temperature. The entropy production gives rise to a loss term $T_k \sigma_{\min}$.

Finally, also q_k must satisfy (2.155). Now due to (2.169), q_k can be expressed in terms of the other quantities as

$$q_k = P - \sum_{i \neq k} q_i = -T_k \left(\sigma_{\min} + \sum_{i=k+1}^{M} \frac{\hat{q}_i}{T_i} + \sum_{i=1}^{k-1} \frac{\breve{q}_i}{T_i}\right). \tag{2.170}$$

Then the condition (2.155) implies the consistency requirement that

$$\sum_{i=1}^{k} \frac{\breve{q}_i}{T_i} + \sum_{i=k+1}^{M} \frac{\hat{q}_i}{T_i} \leq -\sigma_{\min} \leq \sum_{i=1}^{k-1} \frac{\breve{q}_i}{T_i} + \sum_{i=k}^{M} \frac{\hat{q}_i}{T_i} \quad \text{for some } k. \tag{2.171}$$

If the second law constraint (2.157) is fulfilled for all $(q_1,\ldots,q_n)$ satisfying (2.155), then $\tau = 0$. Only the first law of thermodynamics matters, and

$$P = \sum_{i=1}^{M} \hat{q}_i . \tag{2.172}$$

2.4.2 *The maximal power available from a set of stationary flows*

We have solved the problem of finding the maximal amount of exergy flow (mechanical power) from a given set of heat sources and heat sinks with stationary flows, adjustable within certain limits. We shall now generalize this to the case of sources and sinks of any extensive quantity.

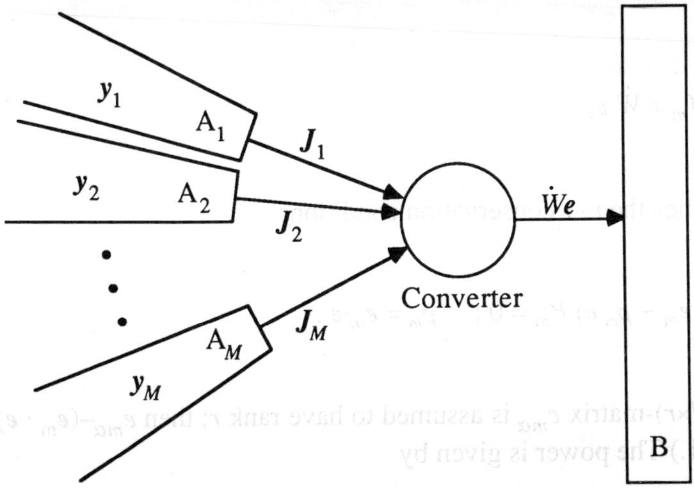

Fig. 2.6. Work extracted out of stationary flows

We consider a set of separate systems $A_1,\ldots, A_M$, each of which can yield (or absorb) a stationary flow of extensive quantities. These flows are then vectors $\boldsymbol{J}_1,\ldots, \boldsymbol{J}_M$ in the r-dimensional space of extensive quantities. We assume that $M \geq r{+}1$. We further assume that each flow carries constant proportions of these quantities, described by unit vectors with non-negative components,

$$J_m = P_m\, e_m\,;\quad e_m \geq 0\,;\quad e_m^2 = 1;\quad (m = 1,\ldots, M), \tag{2.173}$$

where P_m are flow intensity variables. The flows are counted positive going *into* the exchanger (Fig. 2.6). The flow intensities are assumed to be restricted to certain intervals of non-zero widths,

$$\check{P}_m \le P_m \le \hat{P}_m; \;\; \check{P}_m < \hat{P}_m. \tag{2.174}$$

The (fixed) intensive variables of the flows are assumed to be $\boldsymbol{y}_1,..., \boldsymbol{y}_M$ respectively.

The work extracted per unit time (the *power*) into the work reservoir B is assumed to be W$\boldsymbol{e}$, where $\boldsymbol{e}$ is a fixed unit vector characteristic for B, thus

$$\sum_{m=1}^{M} \boldsymbol{J}_m = \dot{W}\,\boldsymbol{e}\,, \tag{2.175}$$

this implies the r–1 conservation conditions

$$\sum_{m=1}^{M} (\boldsymbol{e}_m - \rho_m\,\boldsymbol{e})\,P_m = 0\,; \qquad \rho_m = \boldsymbol{e}_m \cdot \boldsymbol{e}\,. \tag{2.176}$$

(The $(M{\times}r)$-matrix $\boldsymbol{e}_{m\alpha}$ is assumed to have rank r; then $\boldsymbol{e}_{m\alpha}-(\boldsymbol{e}_{\mathrm{m}} \cdot \boldsymbol{e})\boldsymbol{e}_\alpha$ has rank r–1.) The power is given by

$$\dot{W} = \sum_{m=1}^{M} \rho_m\,P_m\,. \tag{2.177}$$

The restrictions (2.174) on the flows are assumed to be compatible with the conservation conditions (2.176).

The *entropy production* $\dot{S}$ must equal the net entropy flow out from the exchanger. The entropy flows are according to (2.3) $-\boldsymbol{y}_m \cdot \boldsymbol{J}_m$, where a factor k has now been included in $\boldsymbol{y}_m$, and since the work does not carry entropy,

$$\dot{S} = -\sum_{m=1}^{M} \boldsymbol{y}_m \cdot \boldsymbol{J}_m = -\sum_{m=1}^{M} v_m\,P_m\,, \tag{2.178}$$

where the specific entropy of each flow is

$$v_m = y_m \cdot e_m > 0. \tag{2.179}$$

The entropy production is assumed to be limited from below by a convex function of the flows P_m connected to the 'forces' necessary to drive the flows through the system,

$$\dot{S} = -\sum_{m=1}^{M} v_m P_m \geq X(\mathbf{P}) ; \quad \mathbf{P} = (P_1, \ldots, P_M). \tag{2.180}$$

The conditions (2.174) also have to be compatible with (2.180).

We are now ready to formulate the problem concisely as: Given $\boldsymbol{e}$; $e_m \geq 0$, y_m, and $\breve{P}_m < \hat{P}_m$ (m=1,..., M), $X(\mathbf{P})$, and

$$\begin{cases} \rho_m = \boldsymbol{e} \cdot \boldsymbol{e}_m , \\ v_m = \boldsymbol{y}_m \cdot \boldsymbol{e}_m > 0, \end{cases} \tag{2.181}$$

maximize the linear form

$$\dot{W} = \sum_{m=1}^{M} \rho_m P_m , \tag{2.177}$$

subject to the constraints (assumed to be mutually consistent)

$$\begin{cases} \breve{P}_m \leq P_m \leq \hat{P}_m ; \\ \sum_{m=1}^{M} (\boldsymbol{e}_m - \rho_m \boldsymbol{e}) P_m = 0; \\ \dot{S} = -\sum_{m=1}^{M} v_m P_m \geq X(\mathbf{P}). \end{cases} \tag{2.182}$$

Since X is a convex function, the set Ω of P's satisfying (2.181) is also convex. Our problem then has a solution on the boundary of Ω, unique except for certain degenerate cases. For the one-dimensional case (r=1) and constant X this problem was solved in the Section 2.4.1.

We now introduce the Lagrangian

$$L = \sum_{m=1}^{M} [\rho_m P_m + \check{\mu}_m (P_m - \check{P}_m) + \hat{\mu}_m (\hat{P}_m - P_m) + $$

$$+ s\cdot(e_m - \rho_m e)\, P_m - \tau v_m P_m] - \tau X(\mathbf{P}), \tag{2.183}$$

where $\check{\mu}_m$, $\hat{\mu}_m$, s and τ are Lagrange multipliers, satisfying

$$\begin{cases} \check{\mu}_m, \hat{\mu}_m, \tau \geq 0; \\ s\cdot e = 0. \end{cases} \tag{2.184}$$

The conditions for maximal L are

$$\frac{\partial L}{\partial P_m} = \check{\mu}_m - \hat{\mu}_m + \mu_m = 0, \tag{2.185}$$

where

$$\begin{cases} \mu_m = \kappa_m - \tau \dfrac{\partial X(\mathbf{P})}{\partial P_m}; \\ \kappa_m = \rho_m - \tau\, v_m + s\cdot e_m; \end{cases} \tag{2.186}$$

$$\begin{cases} \check{\mu}_m \dfrac{\partial L}{\partial \check{\mu}_m} = \check{\mu}_m (P_m - \check{P}_m) = 0; \\ \hat{\mu}_m \dfrac{\partial L}{\partial \hat{\mu}_m} = \hat{\mu}_m (\hat{P}_m - P_m) = 0; \end{cases} \tag{2.187}$$

$$\tau\left(\sum_{m=1}^{M} v_m P_m + X(\mathbf{P})\right) = 0, \tag{2.188}$$

clearly either $\check{\mu}_m$ or $\hat{\mu}_m$ must vanish. When P_m is in the interior of the allowed interval, they both vanish, and then also $\mu_m = 0$. If instead $\mu_m \neq 0$, then P_m must lie at one end of the interval (2.174)

$$P_m = \hat{P}_m\, \theta(\mu_m) + \check{P}_m\, \theta(\mu_m) \quad \text{for } \mu_m \neq 0, \tag{2.189}$$

where θ is the step function

$$\theta(x) = \begin{cases} 1, & x \geq 0; \\ 0, & x < 0. \end{cases} \tag{2.190}$$

Consider now a partition π of the indices m into two sets, A_π and B_π, so that

$$\left\{ \begin{array}{l} \mu_m \neq 0; \\ P_m = \hat{P}_m \quad \text{for } \kappa_m > \tau \dfrac{\partial X}{\partial P_m} \\ \qquad\qquad\qquad\qquad\qquad m \in A_\pi \\ P_m = \check{P}_m \quad \text{for } \kappa_m < \tau \dfrac{\partial X}{\partial P_m} \end{array} \right. \tag{2.191}$$

$$\left\{ \begin{array}{l} \mu_m = \kappa_m - \tau \dfrac{\partial X(\mathbf{P})}{\partial P_m} = 0 \\ \qquad\qquad\qquad\qquad\qquad m \in B_\pi \\ \check{P}_m < P_m < \hat{P}_m \end{array} \right. \tag{2.192}$$

Here we should have at most $M-r+1$ indices in A_π and hence at least $r-1$ indices in B_π. Now (2.192) and (2.186) with (2.181) and (2.184) give for

$$z = s + e \tag{2.193}$$

the following system of equations,

$$\begin{cases} z{\cdot}e = 1; \\ z{\cdot}e_m = \tau\left(v_m + \dfrac{\partial X(\mathbf{P})}{\partial P_m}\right); \quad m \in B_\pi \,. \end{cases} \tag{2.194}$$

We have two main cases according to whether $\tau = 0$, or $\tau > 0$.

Case 1 : $\tau = 0$. To solve (2.194), exactly r–1 vectors $e_m - \rho_m e$ ($m \in B_\pi$) must be linearly independent. After z has been determined μ_m ($m \in A_\pi$) can be obtained from

$$\mu_m = z{\cdot}e_m\,; \quad m \in A_\pi \quad (\tau = 0). \tag{2.195}$$

Then the M–r+1 equations for P_m in (2.189) and the r–1 independent equations (2.176) completely determine P_m (m=1,..., M). $\tau = 0$ is a very special case, in which the region determined by (2.174) does not include any part of the surface corresponding to equality in (2.180). Thus for an increased inflow there are potential improvements to be made through increased efficiency. The marginal work productivity of an inflow may thus be higher than what Carnot factors in (2.200) below would indicate.

Case 2: $\tau > 0$. In this case at most M–r indices belong to A_π and hence at least r indices belong to B_g. Let us assume that there are r+k indices in B_π ($k \geq 0$) and M–r–k indices in A_π. Then there must be k linear relations among the e_m's ($m \in B_g$). Let those relations be

$$\sum_{m \in B_\pi} b_m^{(j)}\, e_m = 0, \quad (j = 1,\ldots, k). \tag{2.196}$$

Then, because of (2.194) the P_m's must satisfy

$$\sum_{m \in B_\pi} b_m^{(j)} \left(v_m + \frac{\partial X(\mathbf{P})}{\partial P_m} \right) = 0, \quad (j = 1,\ldots, k). \tag{2.197}$$

From (2.186) we have

$$\sum_{m=1}^{M} v_m P_m + X(P_1,\ldots, P_M) = 0. \tag{2.198}$$

The k+1 equations (2.197) and (2.198) together with the r–1 equations (2.176) and the M–r–k equations (2.191) form a set of M independent equations which determine P_m completely; let the result be $P_m^{(\pi)}$. We then use the resulting $(\partial X/\partial P_m)^{(\pi)}$ in the system of equations (2.194) to determine z and τ; let the result be $z^{(\pi)}$ and $\tau^{(\pi)}$.

Let the largest value of $W^{(\pi)}$ be W^* and assume that it obtains for the partition $\pi = \pi^*$. Except for certain degenerate cases, π^* is unique. We shall assume this to be the case. Then the maximal power output W^* is obtained for P=P^*, where P^* lies on the boundary of Ω, and moreover on the surface (2.198).

Thus, the extractable power is (star denotes the choice $\pi = \pi^*$).

$$\dot{W}^* = \sum_{m=1}^{M} \rho_m P_m^* = \sum_{m=1}^{M} \kappa_m^* P_m^* - \tau^* X(\mathbf{P}^*), \tag{2.199}$$

where κ_m^* are generalized Carnot factors,

$$\kappa_m^* = \tau^*(y^* - y_m)\cdot e_m\,; \quad (y^* = z^*/\tau^*), \tag{2.200}$$

in which $\boldsymbol{y}^*$ plays the role of reference vector for the intensive variables, and where τ^* defines a reference temperature. With this our problem has been solved.

Exergy may be considered as 'physical value'. The Carnot factors κ_m^* are then values per unit of the various flows, and τ^* is the exergy cost per unit entropy produced.

For applications it is important to determine the function $X(\mathbf{P})$. In the region of linear processes, described by the Onsager relations (2.57), $X(\mathbf{P})$ is quadratic in $\mathbf{P}$, corresponding to losses through "resistance" in the terminology of Andresen et al. [1977].

In this section we have used methods almost parallell to economic theory. In Chapter 7 we shall take one more step towards economics and discuss a production system providing a flow of (physically defined) services from stationary flows, restricted to certain intervals. The nature of such restrictions will also be discussed.

3. GEOMETRICAL INFORMATION THEORY

Many systems use exergy in some form to order matter spatially and to keep the order over a considerable time span. The spatial order then often serves to channel the continued exergy flow.

To describe information manifested in spatial order, we shall combine information theory with geometry in this chapter [Eriksson and Lindgren 1987]. We start with a continuous scalar spatial distribution.

3.1 Spatial Decomposition of Information

Consider a system whose microstate is described by a position vector $\mathbf{x}$ in an n-dimensional Euclidean space E^n and whose macrostate is described by a probability density $p(\mathbf{x})$ in this space,

$$\begin{cases} 0 \le p(\mathbf{x}) < \infty, \quad \mathbf{x} \in E^n; \\ \int d^n\mathbf{x}\, p(\mathbf{x}) = 1, \end{cases} \tag{3.1}$$

where we have chosen $\nu(\mathbf{x}) = 1$ (cf. (1.68)). The entropy of the macrostate described by $p(\mathbf{x})$ is

$$S[p] = \int d^n\mathbf{x}\, p(\mathbf{x}) \ln \frac{1}{p(\mathbf{x})} \, . \tag{3.2}$$

Let $p_0(\mathbf{x})$ be a reference density in E^n, which is non-zero everywhere,

$$\begin{cases} 0 < p_0(\mathbf{x}) < \infty; \\ \int d^n\mathbf{x}\, p_0(\mathbf{x}) = 1. \end{cases} \tag{3.3}$$

Then the Kullback information or contrast of p with respect to p_0 is (cf.

(1.17)),

$$K[p_o; p] = \int d^n\mathbf{x}\, p(\mathbf{x}) \ln \frac{p(\mathbf{x})}{p_o(\mathbf{x})} \geq 0. \tag{3.4}$$

Only if $p(\mathbf{x}) = p_o(\mathbf{x})$, except on a set of measure zero, the contrast vanishes.

We shall analyze the entropy (3.2) and the contrast (3.4) of the probability density (3.1) with respect to the contributions from structures of different size. For a given resolution length,

$$r = e^{-\gamma}, \tag{3.5}$$

where γ indicates the order of magnitude of structure size, we construct an 'observed' probability density $p(\gamma; \mathbf{x})$ from the 'true' density $p(\mathbf{x})$ through convolution with a Gaussian function

$$p(\gamma; \mathbf{x}) = (2\pi)^{-n/2} \int d^n\mathbf{z}\, e^{-\frac{1}{2}\mathbf{z}^2}\, p(\mathbf{x} + e^{-\gamma}\mathbf{z}) =$$

$$= \exp\left[\frac{1}{2} e^{-2\gamma} \nabla^2\right] p(\mathbf{x}). \tag{3.6}$$

We have chosen to call γ the clarity of the observation; then infinite clarity corresponds to perfect resolution.

The density satisfies the differential equation

$$\left(\frac{\partial}{\partial \gamma} + e^{-2\gamma} \nabla^2\right) p(\gamma; \mathbf{x}) = 0, \tag{3.7}$$

and the relations

$$\begin{cases} p(\gamma;\mathbf{x}) > 0; \\ \lim\limits_{\gamma\to\infty} p(\gamma;\mathbf{x}) = p(\mathbf{x}); \\ \int \mathrm{d}^n\mathbf{x}\, p(\gamma;\mathbf{x}) = 1; \\ \int \mathrm{d}^n\mathbf{x}\, p(\gamma;\mathbf{x})\,\mathbf{x} = <\mathbf{x}> \equiv \int \mathrm{d}^n\mathbf{x}\, p(\mathbf{x})\,\mathbf{x}\,; \\ \int \mathrm{d}^n\mathbf{x}\, p(\gamma;\mathbf{x})\, x_j\, x_k = <x_j\, x_k> + \mathrm{e}^{-2\gamma}\,\delta_{jk}\,. \end{cases} \tag{3.8}$$

In the limit $\gamma \rightarrow -\infty$, the two distributions $p(\gamma,\mathbf{x})$ and $p_o(\gamma;\mathbf{x})$ become indistinguishable,

$$\lim_{\gamma\to\infty} \frac{p(\gamma;\mathbf{x})}{p_o(\gamma;\mathbf{x})} = 1. \tag{3.9}$$

We can now use (3.7), (3.8), and (3.9) to rewrite the contrast (3.4) as follows,

$$\begin{aligned} K[p_o;p] &= \int_{-\infty}^{\infty} \mathrm{d}\gamma \frac{\partial}{\partial\gamma} K[p_o;p(\gamma;\cdot)] = \\ &= \int_{-\infty}^{\infty} \mathrm{d}\gamma\, \mathrm{e}^{-2\gamma} \int \mathrm{d}^n\mathbf{x} \left[\frac{p(\gamma;\mathbf{x})}{p_o(\gamma;\mathbf{x})} \nabla^2 p_o(\gamma;\mathbf{x}) \right. \\ &\qquad \left. - \left(1 + \ln \frac{p(\gamma;\mathbf{x})}{p_o(\gamma;\mathbf{x})}\right) \nabla^2 p(\gamma;\mathbf{x}) \right], \end{aligned} \tag{3.10}$$

which may be transformed into

$$\begin{cases} K[p_o; p] = \int_{-\infty}^{\infty} \mathrm{d}\gamma \int \mathrm{d}^n\mathbf{x}\, p(\gamma; \mathbf{x})\, (\mathbf{v}(\gamma; \mathbf{x}))^2; \\ \mathbf{v}(\gamma; \mathbf{x}) = \mathrm{e}^{-\gamma}\, \nabla \ln \dfrac{p(\gamma; \mathbf{x})}{p_o(\gamma; \mathbf{x})}. \end{cases} \tag{3.11}$$

The first equation in (3.11) is a double decomposition of the contrast with respect to position and structure size. The contribution is everywhere positive. We may interpret

$$d(\gamma) = \int \mathrm{d}^n\mathbf{x}\, p(\gamma; \mathbf{x})\, d(\gamma; \mathbf{x})\,; \quad d(\gamma; \mathbf{x}) = n - (\mathbf{v}(\gamma; \mathbf{x}))^2, \tag{3.12}$$

as the dimension (cf. [Mandelbrot 1983]) of the system for a given structure magnitude γ. It is formed as an average over the local dimension $d(\gamma; \mathbf{x})$, which obviously satisfies the inequality

$$d(\gamma; \mathbf{x}) \leq n\,. \tag{3.13}$$

The local dimension may be negative. Where this happens $p(\gamma; \mathbf{x})$, usually behaves like a quickly decreasing Gaussian and the total contribution from areas with negative $d(\gamma, \mathbf{x})$, is small enough for $d(\gamma)$ to satisfy

$$0 \leq d(\gamma) \leq n\,. \tag{3.14}$$

To have all quantities here well-defined, it is enough that $p(\mathbf{x})$ is a Schwarz distribution, since then $p(\gamma; \mathbf{x})$ is finite and differentiable.

3.2 Generalized Information Current

We shall now construct kinetic concepts which may be useful for analysing dynamical systems, where structure size is changing as a result of dynamics. Consider for instance an originally homogenous chemical self-organizing system, which develops a spatial structure. The information is originally tied to the system itself due to its deviation from equilibrium, i.e., structure size = system size. When internal structure is formed, information moves to smaller structure size. Finally it moves into the microscopic domain, where it is lost as entropy production.

The 'space' we shall use for our description is then configuration space $\mathbf{x}$ and resolution $r = e^{-\gamma}$. (We use here r as a variable rather than γ.) We can then rewrite (3.11) as

$$K[p_o; p] = \int_0^\infty \frac{dr}{r} \int d^n\mathbf{x}\, k(r, \mathbf{x}) \tag{3.15}$$

where

$$k(r, \mathbf{x}) = p(\gamma; \mathbf{x}) \left(r \nabla \ln \frac{p(\gamma; \mathbf{x})}{p_o(\gamma; \mathbf{x})} \right)^2. \tag{3.16}$$

We shall generalize this to the case of a chemical system consisting of M components, labeled $i = 1,\ldots, M$. The system is described by the evolution in time t of the relative concentrations $c_i(\mathbf{x}, t) \geq 0$. We assume that, at any point $\mathbf{x}$, they satisfy the normalization,

$$\sum_{i=1}^{M} c_i(\mathbf{x}, t) = 1. \tag{3.17}$$

Assuming ideal gas conditions, the exergy per molecule (see (2.79)) of the system in units of kT is the contrast

$$K(t) = \int d^n\mathbf{x} \sum_{i=1}^{M} c_i(\mathbf{x}, t) \ln \frac{c_i(\mathbf{x}, t)}{c_{i\,o}}, \tag{3.18}$$

where $c_{io} > 0$ are equilibrium concentrations, normalized as in (3.17). Moreover, we assume that for large $|\mathbf{x}|$, $c_i(\mathbf{x}, t)$ approaches c_{io} fast enough for (3.18) to converge.

The dynamics is given by (dot denotes time derivative)

$$\dot{c}_i(\mathbf{x}, t) = D_i \nabla^2 c_i(\mathbf{x}, t) + F_i(c(\mathbf{x}, t)) \quad (i = 1, \ldots, M). \tag{3.19}$$

Here $\boldsymbol{c} = (c_1, \ldots, c_M)$, and D_i is the ith diffusion constant, and F_i denotes the production of the ith component in the chemical reactions. Clearly

$$F_i(c_o) = 0. \tag{3.20}$$

Moreover, to maintain (3.17), we require

$$\sum_{i=1}^{M} F_i(c) = 0. \tag{3.21}$$

To study structures on different scales, it is convenient to introduce concentrations corresponding to a resolution length r (see (3.6)),

$$c_i(r, \mathbf{x}, t) = (2\pi)^{-n/2} \int \mathrm{d}^n\mathbf{z}\, \mathrm{e}^{-\frac{1}{2}\mathbf{z}^2}\, c_i(\mathbf{x} + r\,\mathbf{z}, t) = \mathrm{e}^{\frac{r^2}{2}\nabla^2}\, c_i(\mathbf{x}, t), \tag{3.22}$$

which also satisfy the normalization (3.17), and the differential equation (cf. (3.7))

$$\left(\frac{1}{r}\frac{\partial}{\partial r} - \nabla^2\right) c_i(r, \mathbf{x}, t) = 0. \tag{3.23}$$

The equation of motion for $c_i(r, \mathbf{x}, t)$ is easily obtained from (3.19),

$$\dot{c}_i(r, \mathbf{x}, t) = D_i \nabla^2 c_i(r, \mathbf{x}, t) + \mathrm{e}^{\frac{r^2}{2}\nabla^2} F_i(\boldsymbol{c}(\mathbf{x}, t)). \tag{3.24}$$

The contrast (3.18) may be decomposed similarly to (3.15) and (3.16),

$$K(t) = \int_0^\infty \frac{\mathrm{d}r}{r} \int_{-\infty}^{\infty} \mathrm{d}x\, k(r, \mathbf{x}, t), \tag{3.25}$$

where the positive semi-definite information (or contrast) density in space

$\mathbf{x}$ and scale, $\ln r$, is

$$k(r, \mathbf{x}, t) = r^2 \sum_{i=1}^{M} \frac{(\nabla c_i(r, \mathbf{x}, t))^2}{c_i(r, \mathbf{x}, t)} =$$

$$= \left(-r\frac{\partial}{\partial r} + r^2\,\nabla^2\right) \sum_{i=1}^{M} c_i(r, \mathbf{x}, t) \ln \frac{c_i(r, \mathbf{x}, t)}{c_{i\,\mathrm{o}}}. \tag{3.26}$$

The entropy (in units of Bolzmann's constant k) production density (in $\mathbf{x}$) connected to (3.19) is

$$\sigma(\mathbf{x}, t) = \sum_{i=1}^{M} \left[D_i \frac{(\nabla c_i(\mathbf{x}, t))^2}{c_i(\mathbf{x}, t)} - \left(\ln \frac{c_i(\mathbf{x}, t)}{c_{i\,\mathrm{o}}}\right) F_i(\mathbf{c}(\mathbf{x}, t)) \right]. \tag{3.27}$$

The first term here is due to diffusion ($-D_i \nabla c_i(\mathbf{x}, t)$ is the diffusion current density and $c_i(\mathbf{x}, t)^{-1} \nabla c_i(\mathbf{x}, t)$ is the gradient of the chemical potential) and the second term to chemical reactions.

We identify $\sigma(\mathbf{x}, t)$ with the scale component of the contrast current density, $j_r(r, \mathbf{x}, t)$, disappering into the microscopic scale ($r = 0$),

$$j_r(0, \mathbf{x}, t) = \sigma(\mathbf{x}, t). \tag{3.28}$$

In analogy with this, for a finite resolution r we define

$$j_r(r, \mathbf{x}, t) = \sum_{i=1}^{M} \left[D_i \frac{(\nabla c_i(r, \mathbf{x}, t))^2}{c_i(r, \mathbf{x}, t)} \right.$$

$$\left. - \left(\ln \frac{c_i(r, \mathbf{x}, t)}{c_{i\,\mathrm{o}}}\right) e^{\frac{r^2}{2}\nabla^2} F_i(\mathbf{c}(\mathbf{x}, t)) \right]. \tag{3.29}$$

To determine the spatial part $\mathbf{j}(r, \mathbf{x}, t)$ of the contrast current density we require k and $(j_r, \mathbf{j})$ to satisfy a continuity equation,

$$\dot{k}(r, \mathbf{x}, t) = r \frac{\partial}{\partial r} j_r(r, \mathbf{x}, t) - \nabla \cdot \mathbf{j}(r, \mathbf{x}, t). \tag{3.30}$$

We can now use this together with (3.29) and (3.26) and the equation of motion (3.24) to determine $\nabla \cdot \mathbf{j}(r, \mathbf{x}, t)$. The result is

$$\nabla \cdot \mathbf{j}(r, \mathbf{x}, t) = - r^2 \nabla^2 \sum_{i=1}^{M} \left(\ln \frac{c_i(r, \mathbf{x}, t)}{c_{io}} \right) e^{\frac{r^2}{2}\nabla^2} F_i(\mathbf{c}(\mathbf{x}, t)). \tag{3.31}$$

Requiring $\mathbf{j}(r, \mathbf{x}, t)$ to disappear at large distances where $c_i(r, \mathbf{x}, t)$ is close to c_{io} and to be rotation-free, we immediately conclude that

$$\mathbf{j}(r, \mathbf{x}, t) = - r^2 \nabla \sum_{i=1}^{M} \left(\ln \frac{c_i(r, \mathbf{x}, t)}{c_{io}} \right) e^{\frac{r^2}{2}\nabla^2} F_i(\mathbf{c}(\mathbf{x}, t)). \tag{3.32}$$

We thus have for the contrast density and current the result

$$\begin{cases} k(r, \mathbf{x}, t) = r^2 \displaystyle\sum_{i=1}^{M} \frac{(\nabla c_i(r, \mathbf{x}, t))^2}{c_i(r, \mathbf{x}, t)} ; \\ \mathbf{j}(r, \mathbf{x}, t) = - r^2 \nabla \displaystyle\sum_{i=1}^{M} \left(\ln \frac{c_i(r, \mathbf{x}, t)}{c_{io}} \right) e^{\frac{r^2}{2}\nabla^2} F_i(\mathbf{c}(\mathbf{x}, t)) ; \\ j_r(r, \mathbf{x}, t) = \displaystyle\sum_{i=1}^{M} \left[D_i \frac{(\nabla c_i(r, \mathbf{x}, t))^2}{c_i(r, \mathbf{x}, t)} - \left(\ln \frac{c_i(r, \mathbf{x}, t)}{c_{io}} \right) e^{\frac{r^2}{2}\nabla^2} F_i(\mathbf{c}(\mathbf{x}, t)) \right]. \end{cases} \tag{3.33}$$

We then integrate over scale to find the ordinary density and current,

$$\begin{cases} k(\mathbf{x},t) = \int_0^\infty \frac{\mathrm{d}r}{r}\, k(r,\mathbf{x},t) = \sum_{i=1}^{M} c_i(\mathbf{x},t) \ln \frac{c_i(\mathbf{x},t)}{c_{i\,0}} + \\ \qquad\qquad + \nabla^2 \int_0^\infty \mathrm{d}r\, r \sum_{i=1}^{M} c_i(r,\mathbf{x},t) \ln \frac{c_i(r,\mathbf{x},t)}{c_{i\,0}} > 0; \\ \mathbf{j}(\mathbf{x},t) = \int_0^\infty \frac{\mathrm{d}r}{r}\, \mathbf{j}(r,\mathbf{x},t) = \sum_{i=1}^{M} (-D_i \nabla c_i(\mathbf{x},t)) \ln \frac{c_i(\mathbf{x},t)}{c_{i\,0}} \\ \qquad\qquad - \frac{\partial}{\partial t} \nabla \int_0^\infty \mathrm{d}r\, r \sum_{i=1}^{M} c_i(r,\mathbf{x},t) \ln \frac{c_i(r,\mathbf{x},t)}{c_{i\,0}}, \end{cases} \tag{3.34}$$

where we have used (3.26) and (3.24). From (3.33), (3.30), and (3.28) we then get the balance equation for contrast,

$$\dot{k}(\mathbf{x},t) + \nabla\cdot\mathbf{j}(\mathbf{x},t) + \sigma(\mathbf{x},t) = 0. \tag{3.35}$$

In self-organizing systems information is moving in scale, passing from the

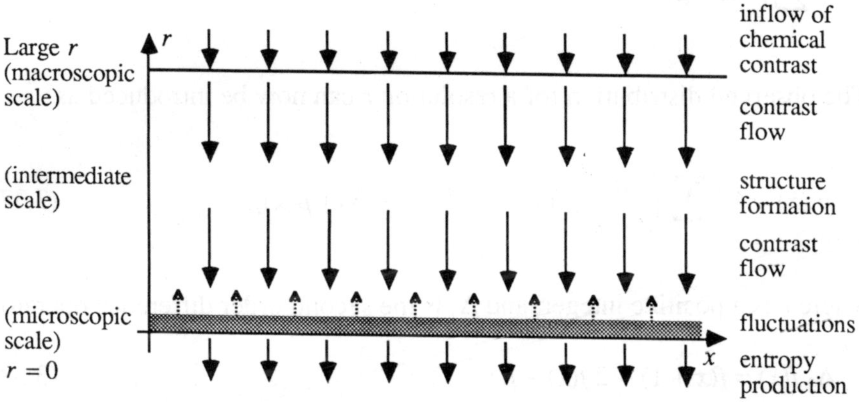

Fig. 3.1. Diagram, indicating the flow of information from the macroscopic scale (chemical contrast) to the microscopic scale, via an intermediate scale with structure formation triggered by fluctuations.

macroscopic scale (chemical exergy) to the microscopic scale, and halting along the way at an intermediate scale (entropy production), as indicated in Fig. 3.1. The information flow from fluctuations going in the other direction will be discussed below in Section 3.4.2. The concepts introduced here are designed to be useful in the analysis of such processes. In Section 5.2 they will be applied to a model of a chemical self-organizing system.

3.3 Structural, Correlational, and Textual Contrast

In this section information-theoretical concepts are introduced, for a one-dimensional lattice, to be used to characterize the evolving structure in models of self-organizing systems. Some concepts are more easily defined for a lattice than in the continuum formulation of the previous section.

Consider for simplicity an infinite one-dimensional lattice, where each site is characterized by an integer x, and a normalized non-negative probability distribution $p(x)$.

$$\begin{cases} p(x) \geq 0 \qquad (x = \ldots, -2, -1, 0, 1, 2, \ldots); \\ \sum_x p(x) = 1. \end{cases} \tag{3.36}$$

The observed distribution for a resolution r can now be introduced as

$$p_r(x) = 4^{-r} \sum_{k=-r}^{r} \binom{2r}{r+k} p(x+k) = \left(1 + \tfrac{1}{4}\,\Delta_2\right)^r p(x), \tag{3.37}$$

where r is a positive integer, and Δ_2 is the second order difference operator

$$\Delta_2 f(x) = f(x+1) - 2f(x) + f(x-1). \tag{3.38}$$

Then $p_r(x)$ satisfies the conditions corresponding to (3.8),

$$\begin{cases} p_r(x) \geq 0, \quad r \geq 1; \\ p_o(x) = p(x); \\ \sum_x p_r(x) = 1; \\ \sum_x p_r(x)\, x = \langle x \rangle \equiv \sum_x p(x)\, x\,; \\ \sum_x p_r(x)\, x^2 = \langle x^2 \rangle + \frac{1}{2} r\,. \end{cases} \tag{3.39}$$

where the last equation shows that $r/2$ now plays the role of r^2 in (3.5-8). The binomial lattice distribution used in (3.37)

$$\begin{cases} Q^{(r)} = \{q_r(x)\}_{-r}^{r}\,; \qquad q_r(x) = 2^{-2r}\binom{2r}{r+x}; & (3.40) \\ \sum_{x=-r}^{r} q_r(x)\, x^2 = \frac{1}{2}\, r & (3.41) \end{cases}$$

corresponds to the Gaussian

$$q(r;x) = \frac{1}{\sqrt{\pi r}}\, \mathrm{e}^{-x^2/r}\,; \qquad \int_{-\infty}^{\infty} \mathrm{d}x\, q(r;x)\, x^2 = \frac{1}{2}\, r \tag{3.42}$$

in the continuous case. The entropy of the discrete distribution (3.40) $S[Q^{(r)}]$ converges quite fast towards that of the continuous one (cf. (1.80)),

$$S[q(r;\cdot)] = \frac{1}{2}\,(1 + \ln(\pi r)). \tag{3.43}$$

Let now $p^{(o)}(x)$ be a reference distribution, which is non-zero for each x, for which $p(x)$ is non-zero. The contrast

$$K[p^{(o)}; p] = \sum_x p(x) \ln \frac{p(x)}{p^{(o)}(x)} \tag{3.44}$$

may then be decomposed as follows,

$$K[p^{(o)}; p] = \sum_{r=0}^{\infty} |\Delta K_r| \tag{3.45}$$

with

$$\begin{aligned}
|\Delta K_r| &= K[p_r^{(o)}; p_r] - K[p_{r+1}^{(o)}; p_{r+1}] = \\
&= \sum_x \left[p_r(x) \ln \frac{p_r(x)}{p_r^{(o)}(x)} - p_{r+1}(x) \ln \frac{p_{r+1}(x)}{p_{r+1}^{(o)}(x)} \right] = \\
&= \sum_x p_{r+1}(x) \sum_{\varepsilon_1, \varepsilon_2 = \pm\frac{1}{2}} \frac{\frac{1}{4} p_r(x + \varepsilon_1 + \varepsilon_2)}{p_{r+1}(x)} \\
&\quad \times \ln \frac{\frac{1}{4} p_r(x + \varepsilon_1 + \varepsilon_2)/p_{r+1}(x)}{\frac{1}{4} p_r^{(o)}(x + \varepsilon_1 + \varepsilon_2)/p_{r+1}^{(o)}(x)} = \\
&= \sum_x p_{r+1}(x)\, K[\hat{c}_r^{(o)}(x); \hat{c}_r(x)],
\end{aligned} \tag{3.46}$$

where

$$\hat{c}_r(x; \varepsilon_1, \varepsilon_2) = \frac{p_r(x + \varepsilon_1 + \varepsilon_2)}{4p_{r+1}(x)}, \quad \varepsilon_1, \varepsilon_2 = \pm\frac{1}{2}; \tag{3.47}$$

and similarly for $\hat{c}_r{}^{(o)}(x; \varepsilon_1, \varepsilon_2)$. Thus,

$$\begin{cases} K[p^{(o)}; p] = \sum_{r=0}^{\infty} \sum_x k_r(x); \\ k_r(x) = p_{r+1}(x)\, K[\hat{c}_r^{(o)}(x); \hat{c}_r(x)], \end{cases} \tag{3.48}$$

in analogy to (3.11). The expressions for the contrast dimensions, similar to (3.12), are

$$\begin{cases} d(r) = 1 - 2\,r\,|\,\Delta K_r\,| = \sum_x p_{r+1}(x)\; d(r;x); \\ d(r;x) = 1 - 2\,r\,K[\hat{c}_r^{(o)}(x);\,\hat{c}_r(x)], \end{cases} \tag{3.49}$$

where the factors $2r$ come from the fact that with the present definition of resolution r, the clarity is

$$\gamma = \frac{1}{2}\ln\frac{r}{2}\,. \tag{3.50}$$

3.3.1 *Chemical system on a one-dimensional lattice*

Consider again an infinite one-dimensional lattice, where at each point x we have a normalized probability distribution $c_i(x)$ over M available chemical species ($i = 1,\dots, M$),

$$\begin{cases} c_i(x) \geq 0, \quad \text{all } x \quad (i = 1,\dots, M); \\ \sum_{i=1}^{M} c_i(x) = 1. \end{cases} \tag{3.51}$$

We shall think of $c_i(x)$ as relative concentrations of the M species.

We assume that the system is large-scale homogeneous, so that the correlation function for a sequence of m neighbouring sites

$$q^{(m)}_{i_1\dots i_m} = \lim_{L\to\infty} \frac{1}{2L+1} \sum_{k=-L}^{L} c_{i_1}(x+k)\,\dots\, c_{i_m}(x+m-1+k) \tag{3.52}$$

exists for any finite m. For $m = 1$ this reduces to the average concentrations

$$\bar{c}_i = q_i^{(1)}. \tag{3.53}$$

The q's (3.52) satisfy the sum rules,

$$\sum_{j=1}^{M} q^{(m)}_{i_1 \dots i_{m-1} j} = \sum_{j=1}^{M} q^{(m)}_{j i_1 \dots i_{m-1}} = q^{(m-1)}_{i_1 \dots i_{m-1}} . \tag{3.54}$$

They may or may not be symmetric under space inversion

$$x \to -x \Rightarrow q^{(m)}_{i_1 \dots i_m} \to q^{(m)}_{i_m \dots i_1} . \tag{3.55}$$

In a chemical system we may often expect symmetry under (3.55). On the other hand, if the lattice is the set of positions of letters in a text, there is usually no symmetry under (3.55).

If the concentrations for chemical equilibrium are

$$\begin{cases} c_i^{(0)}(x) = c_i^{(0)} > 0, \quad \text{all } x \quad (i = 1, \dots, M); \\ \sum_{i=1}^{M} c_i^{(0)} = 1, \end{cases} \tag{3.56}$$

then, in units of kT, the exergy in an interval $I = [a, a+L-1]$ of the lattice (containing L lattice points and N particles), is given by the particle density N/L times the total contrast $K(I)$ for the interval, where

$$K(I) = \sum_{x=a}^{a+L-1} K[c^{(0)}; c(x)] = \sum_{x=a}^{a+L-1} \sum_{i=1}^{M} c_i(x) \ln \frac{c_i(x)}{c_i^{(0)}} . \tag{3.57}$$

For a sufficiently large L we can decompose this into a term based on the mean concentrations (3.53), and a term which contains all the spatial dependence,

$$K(I) = K_{\text{chem}}(I) + K_{\text{str}}(I), \tag{3.58}$$

where

$$\begin{cases} K_{\text{chem}}(I) = L\, k_{\text{chem}}\,; & (3.59) \\ K_{\text{str}}(I) = L\, S[\bar{c}\,] - \sum_{x=a}^{a+L-1} S[c(x)] = \sum_{x=a}^{a+L-1} k_{\text{str}}(x) & (3.60) \end{cases}$$

are the *chemical* and *structural contrasts*, and k_{chem} and $k_{\text{str}}(x)$ are the corresponding quantities per cell,

$$k_{\text{chem}} = \sum_{i=1}^{M} \bar{c}_i \ln \frac{\bar{c}_i}{c_i^{(0)}} = K[c^{(0)};\, \bar{c}]; \tag{3.61}$$

$$\sum_{x=a}^{a+L-1} k_{\text{str}}(x) = \sum_{x=a}^{a+L-1} \sum_{i=1}^{M} c_i(x) \ln \frac{c_i(x)}{\bar{c}_i} = \sum_{x=a}^{a+L-1} K[\,\bar{c};\, c(x)]. \tag{3.62}$$

We see immediately that for $\bar{c}_i = c_i^{(0)}$, the chemical contrast vanishes, and for a spatially homogeneous system, $c_i(x) = \bar{c}_i$, the structural contrast vanishes. We shall decide on the definition of $k_{\text{str}}(x)$ later.

Structural dimension

It is possible to decompose $K_{\text{str}}(I)$ as in (3.48). In terms of the spatial distributions for the different species,

$$\begin{cases} p^{(i)}(x) = \dfrac{c_i(x)}{L\,\bar{c}_i}\,, \quad x \in I\,; \\ p^{(i)}(x) = 0, \quad x \notin I\,; \end{cases} \qquad \begin{cases} \displaystyle\sum_{x=a}^{a+L-1} p^{(i)}(x) = 1; \\ \displaystyle\sum_{i=1}^{M} \bar{c}_i\, p^{(i)}(x) = \frac{1}{L} \end{cases} \tag{3.63}$$

and

$$\begin{cases} p^{(i,\mathrm{o})}(x) = \dfrac{1}{L}\,, \quad x \in I\,; & \displaystyle\sum_x p^{(i,\mathrm{o})}(x) = 1; \\ p^{(i,\mathrm{o})}(x) = 0\,, \quad x \notin I\,, & \end{cases} \tag{3.64}$$

we can rewrite (3.60) as

$$K_{\mathrm{str}}(I) = \sum_{i=1}^{M} L\,\bar{c}_i\,K_{\mathrm{str}}^{(i)}\,, \tag{3.65}$$

where

$$K_{\mathrm{str}}^{(i)} = K[p^{(i,\mathrm{o})};\,p^{(i)}] \tag{3.66}$$

is the structural contrast per cell for the ith component. If the structural contrast is decomposed as in (3.45) and (3.46).

$$K_{\mathrm{str}}^{(i)} = \sum_{r=0}^{\infty} |\,\Delta K_r^{(i)}\,|\,, \tag{3.67}$$

the contributions from $r > 0$ can be expressed in terms of structual contrast dimension (3.49),

$$|\,\Delta K_r^{(i)}\,| = \frac{1 - d_{\mathrm{str}}^{(i)}(r)}{2r}\,, \quad r > 0, \tag{3.68}$$

where

$$d_{\mathrm{str}}^{(i)}(r) = \sum_x p_{r+1}^{(i)}(x)\;d_{\mathrm{str}}^{(i)}(r;\,x). \tag{3.69}$$

The observed spatial distribution for a resolution $r + 1$ was defined in (3.37). The local structural contrast dimension for the ith species is defined as

$$d_{\mathrm{str}}^{(i)}(r;x) = 1 - 2\,r\,K[\hat{c}_r^{(i,\,\mathrm{o})}(x);\,\hat{c}_r^{(i)}(x)], \tag{3.70}$$

where $\hat{c}_r^{(i)}(x)$ and $\hat{c}_r^{(i,\mathrm{o})}(x)$ are defined in analogy to (3.47).

One can define structural contrast dimensions, which are averages over the species,

$$\begin{cases} d_{\mathrm{str}}(r;x) = L \sum_{i=1}^{M} \bar{c}_i\, p_{r+1}^{(i)}(x)\, d_{\mathrm{str}}^{(i)}(r;x); \\ d_{\mathrm{str}}(r) = \sum_{i=1}^{M} \bar{c}_i\, d_{\mathrm{str}}^{(i)}(r) = \frac{1}{L} \sum_{x=a}^{a+L-1} d_{\mathrm{str}}(r;x), \end{cases} \tag{3.71}$$

where we have used (3.63) and (3.37). From (3.65) and (3.67) we then find

$$K_{\mathrm{str}}(I) = L \sum_{r=0}^{\infty} \sum_{i=1}^{M} \bar{c}_i\, |\Delta K_r^{(i)}|, \tag{3.72}$$

and

$$\sum_{i=1}^{M} \bar{c}_i\, |\Delta K_r^{(i)}| = \frac{1 - d_{\mathrm{str}}(r)}{2r}, \quad r > 0. \tag{3.73}$$

We have simulated the build-up of structural contrast in a one-dimensional model of a chemical self-organizing system, and the result will be shown below in Section 5.2.

Correlational and textual contrast

To extract from $K_{\mathrm{str}}(x)$ the information content inherent in the correlation functions (3.52), we first define

$$S_m = S[q^{(m)}] = \sum_{i_1,\ldots,\,i_m} q_{i_1\ldots i_m}^{(m)} \ln \frac{1}{q_{i_1\ldots i_m}^{(m)}} \quad (m \geq 1), \tag{3.74}$$

and

$$\begin{cases} \Delta S_m = S_m - S_{m-1}\ ; \ S_o = 0; \\ \Delta^2 S_m = S_m - 2\, S_{m-1} + S_{m-2} \quad (m \geq 2). \end{cases} \tag{3.75}$$

We then define the *correlational* part of the structual contrast as

$$K_{\text{corr}}(l) = \sum_{i_1,\ldots,\, i_L} q^{(L)}_{i_1\ldots i_L} \ln \frac{q^{(L)}_{i_1\ldots i_L}}{\bar{c}_{i_1} \ldots \bar{c}_{i_L}} = G[q^{(L)}] = L\, S_1 - S_L =$$

$$= - \sum_{m=1}^{L-1} (L - m)\, \Delta^2 S_{m+1}\,. \tag{3.76}$$

Here $G[p^{(m)}]$ is the mutual information (1.26) of m systems described by the common probability distribution

$$p^{(m)}_{i_1\ldots i_m} \geq 0\ ; \ \sum_{i_1=1}^{M} \ldots \sum_{i_m=1}^{M} p^{(m)}_{i_1\ldots i_m} = 1, \tag{3.77}$$

where i_k labels the microstate of the kth system. From (3.77) we can sum to obtain single-system distributions

$$p_i(k) = \sum_{i_1,\ldots,\, i_m} \delta_{i_k i}\, p^{(m)}_{i_1\ldots i_m} \qquad (k = 1,\ldots, m). \tag{3.78}$$

The mutual information of the m systems is then

$$G[p^{(m)}] = K[\prod_{k=1}^{m} p(k);\, p^{(m)}] = \sum_{k=1}^{m} (S[p(k)] - S[p^{(m)}]) \geq 0. \tag{3.79}$$

We now assume that correlations saturate at a correlation length L_c much

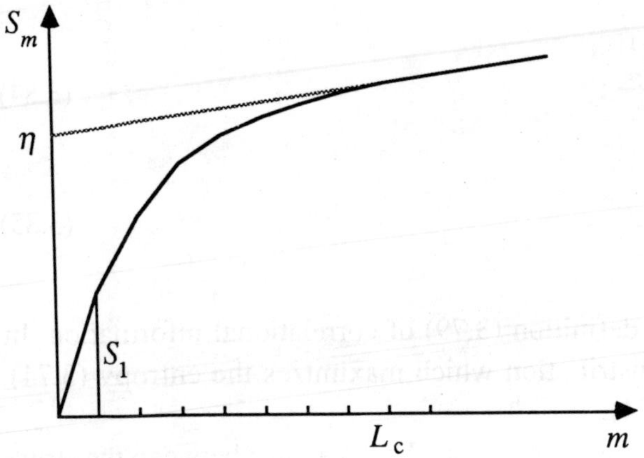

Fig. 3.2. The m-point entropy S_m as a function of correlation length m. For m greater than the maximal correlation length L_c, the entropy grows linearly in m. The intersection of this line with the S_m-axis, η, can be used as a measure for complexity, as will be discussed in Section 4.1.

smaller than the length L of the interval (see Fig. 3.2),

$$\Delta^2 S_m = 0\ , \quad \text{for } m > L_c; \quad L_c << L. \tag{3.80}$$

Neglecting terms of the order L_c/L, we then obtain from (3.76)

$$K_{\text{corr}}(I) = L\, k_{\text{corr}} = L \sum_{m=2}^{L_c} k_{\text{corr}}^{(m)}\ , \tag{3.81}$$

where (see Fig. 3.2)

$$k_{\text{corr}} = S_1 - \Delta S_{L_c}\ , \tag{3.82}$$

and

$$\begin{aligned} k_{\text{corr}}^{(m)} &= -\Delta^2 S_m = 2S_{m-1} - S_m - S_{m-2} = \\ &= \sum_{i_1,\ldots,i_m} q_{i_1\ldots i_m}^{(m)} \ln \frac{q_{i_1\ldots i_m}^{(m)}}{\tilde{q}_{i_1\ldots i_m}^{(m)}} = K[\tilde{q}^{(m)}; q^{(m)}] = \\ &= \sum_{i_2,\ldots,i_{m-1}} q_{i_2\ldots i_{m-1}}^{(m-2)}\, G[t^{(i_2\ldots i_{m-1})}]\ , \end{aligned} \tag{3.83}$$

with

$$\tilde{q}^{(m)}_{i_1\ldots i_m} = \frac{q^{(m-1)}_{i_1\ldots i_{m-1}}\, q^{(m-1)}_{i_2\ldots i_m}}{q^{(m-2)}_{i_2\ldots i_{m-1}}} ; \tag{3.84}$$

$$t^{(i_2\ldots i_{m-1})}_{jj'} = \frac{q^{(m)}_{ji_2\ldots i_{m-1}j'}}{q^{(m-2)}_{i_2\ldots i_{m-1}}} , \tag{3.85}$$

where we have used the definition (3.79) of correlational information. In fact $\tilde{q}^{(m)}$ is the m-point distribution which maximizes the entropy (3.74) under the constraints (3.54).

We now define the textual contrast as the difference between the structural contrast(3.60) and the correlational contrast (3.81) and (3.82),

$$K_{\mathrm{tx}}(I) = K_{\mathrm{str}}(I) - K_{\mathrm{corr}}(I) = L\, \Delta S_{L_c} - \sum_{x=a}^{a+L-1} S[c(x)]. \tag{3.86}$$

Using (3.74) and (3.52) we may rewrite this as

$$K_{\mathrm{tx}}(I) = \sum_{x=a}^{a+L-1} k_{\mathrm{tx}}(x) \tag{3.87}$$

with

$$k_{\mathrm{tx}}(x) = \sum_{i_1,\ldots,\, i_{L_c-1}} c_{i_1}(x - L_c + 1)\ldots c_{i_{L_c-1}}(x-1)\, K[\tilde{c}^{(i_1\ldots i_{L_c-1})};\, c(x)] \tag{3.88}$$

where

$$\tilde{c}^{(i_1\ldots i_{L_c-1})}_i = \frac{q^{(L_c)}_{i_1\ldots i_{L_c-1}i}}{q^{(L_c-1)}_{i_1\ldots i_{L_c-1}}} \tag{3.89}$$

is the conditional probability distribution for a certain cell, given that the $L_c - 1$ preceding cells are occupied by the sequence $i_1,\ldots, i_{L_c-1}$. Then (3.88) expresses the information gain (properly weighted over preceding sequences) when the distribution $c_i(x)$ at x is made known.

Now there are three possible ways of defining non-negative $k_{\mathrm{str}}(x)$ appearing in (3.60). The first is term-by-term identification with (3.62); the second is obtained via (3.66) and (3.67) from term-by-term identification with (3.46). The third is obtained from (3.86) by choosing $k_{\mathrm{str}}(x)$ to be $k_{\mathrm{corr}} + k_{\mathrm{tx}}(x)$, given in (3.82) and (3.88). Thus

$$\begin{cases} k^{\mathrm{I}}_{\mathrm{str}}(x) = K[\bar{c};\, c(x)]; & (3.90) \\ k^{\mathrm{II}}_{\mathrm{str}}(x) = \displaystyle\sum_{r=0}^{\infty}\sum_{i=1}^{M} L\, \bar{c}_i\, p^{(i)}_{r+1}(x)\, K[\hat{c}^{(i,\,o)}_r(x);\, \hat{c}^{(i)}_r(x)]; & (3.91) \\ k^{\mathrm{III}}_{\mathrm{str}}(x) = k_{\mathrm{corr}} + k_{\mathrm{tx}}(x). & (3.92) \end{cases}$$

The contrast in (3.90) is without reference to the neighbouring sites, apart from the mean concentrations $\bar{c}_i$. The contrast in (3.91) is a sum over non-local distributions corresponding to varying resolution; this is the quantity corresponding to $k(x, t)$ in (3.34) for the continuous case. Finally, the contrast in (3.92) is defined to be readily separable into a correlational and a textual part. Note that the contrasts defined in (3.90-92) all have the same spatial average.

The entropy difference ΔS_{L_c} (cf. (3.74) and (3.75)) plays an important role here. It appears in the expressions for correlational and textual contrast (3.82) and (3.86). It has been defined only for a one-dimensional lattice. However, ΔS_{L_c} is quite a simple concept, and it may be generalized to a continuous system, and further to a system of any dimension [Eriksson 1986]. This opens a possibility for a general separation of structural contrast into correlational and textual contrast. In fact, this can be combined with a decomposition with respect to resolution.

3.4 Noise Sensitivity of Dynamical Systems

The influence of noise plays a crucial role in most self-organizing systems. In Section 3.2 we discussed the flow of contrast (information) from macroscale to microscale. We shall now turn to the reverse process, the information generated by microscopic fluctuations, which may be amplified to show up at the macroscopic level.

In spatially homogenous systems, e.g., systems showing periodic or chaotic oscillations, the flow of information from microscale to macroscale is well studied. As a measure for the amplification of small perturbations one uses the Lyapunov exponents, which are ensemble or time averages that show how nearby trajectories diverge in different directions [Oseledec 1968, Shaw 1981, Haken 1984, Eckmann and Ruelle 1985].

In a spatially self-organizing system the contrast flow from microlevel to macrolevel decides which of all possible spatial structures that shall be developed. There is a continous "creation" of information by the fluctuations, and fluctuations in suitable 'directions' in the function space of possible spatial structures can be amplified and contribute to the contrast flow from microlevel to macrolevel.

3.4.1 *Lyapunov exponents*

Consider a spatially homogenous dynamical system where the state $\boldsymbol{x}$ is a vector in R^n, governed by a set of coupled differential equations

$$\dot{\boldsymbol{x}}(t) = \boldsymbol{F}(\boldsymbol{x}(t)). \tag{3.93}$$

To analyze the sensitivity of small perturbations (3.93) is applied to two close trajectories X and X′, such that $\boldsymbol{x}(t) \in \mathrm{X}$ and $\boldsymbol{x}(t) + \boldsymbol{u}(t) \in \mathrm{X}'$, with $|\boldsymbol{u}(t)| << |\boldsymbol{x}(t)|$. Linearization around the trajectory $\boldsymbol{x}(t)$, yields the equation of motion for $\boldsymbol{u}$

$$\dot{\boldsymbol{u}}(t) = \mathbf{J}(\boldsymbol{x}(t))\, \boldsymbol{u}(t). \tag{3.94}$$

Here $\mathbf{J}(\boldsymbol{x})$ is the Jacobian matrix with elements $J_{ij} = \partial F_i/\partial x_j$. Then, for almost all initial conditions $\boldsymbol{u}(0)$, the following limit converges to the maximal Lyapunov exponent λ_1

$$\lim_{t \to \infty} \frac{1}{t} \ln |\boldsymbol{u}(t)| = \lambda_1 . \tag{3.95}$$

This means that, in average, close trajectories diverge in time as $\exp(\lambda_1 t)$.

The Lyapunov exponents can be defined in terms of a time-averaged Jacobian matrix Λ as follows (for further details see, e.g., [Eckman and Ruelle 1985]). Let $\boldsymbol{F}(t)$ denote the mapping from $\boldsymbol{x}(0)$ to $\boldsymbol{x}(t)$, and $\mathbf{J}(t)$ the corresponding Jacobian matrix with the elements $J_{ij}(t) = \partial F_i(t)/\partial x_j$. Then the matrix Λ is defined as

$$\Lambda = \lim_{t \to \infty} (\mathbf{J}(t))^{1/t} \tag{3.96}$$

and the Lyapunov exponents $\lambda_1 > \lambda_2 > \ldots$ are defined as the logarithms of the eigenvalues of Λ, which may be expressed as

$$\lim_{t \to \infty} \frac{1}{t} \ln |\mathbf{J}(t)\boldsymbol{u}| = \lambda_i \ , \quad \text{if } \boldsymbol{u} = \boldsymbol{u}(0) \in \mathrm{E}_i \setminus \mathrm{E}_{i+1} , \tag{3.97}$$

where E_i is the subspace of R^n which corresponds to the eigenvalues that are smaller than $\exp(\lambda_i)$.

As an illustration we may consider a three-dimensional dynamical system. If all Lyapunov exponents are less than zero ($\lambda_{1,2,3} < 0$) the system approaches a stable fix point. If $\lambda_1 = 0$ while $\lambda_{2,3} < 0$ the trajectory tends to a periodic limit cycle. When $\lambda_1 = \lambda_2 = 0$ and $\lambda_3 < 0$, $\boldsymbol{x}(t)$ lies on a stable torus. The system is chaotic, i.e., $\boldsymbol{x}(t)$ approaches a *strange attractor*, if $\lambda_1 > 0$, $\lambda_2 = 0$ and $\lambda_3 < 0$.

In studies of the noise sensitivity in oscillating systems, concepts which quantify local sensitivity have been introduced [Nicolis et al. 1983, Fujisaka 1983, Grassberger and Procaccia 1984, Herzel et al. 1987]. The logarithmic time derivative of the Euclidean norm $|\boldsymbol{u}(t)|$ can be written

$$\frac{\mathrm{d}}{\mathrm{d}t}\ln|\boldsymbol{u}(t)| = L(\boldsymbol{x}(t), \boldsymbol{u}(t)), \tag{3.98}$$

where

$$L(\boldsymbol{x}, \boldsymbol{u}) = \frac{\boldsymbol{u}\cdot(\mathbf{J}(\boldsymbol{x})\,\boldsymbol{u})}{\boldsymbol{u}^2} \tag{3.99}$$

defines a divergence rate which depends on $\boldsymbol{x}(t)$ and the direction of $\boldsymbol{u}(t)$. Since, for almost all $\boldsymbol{u}(0)$, $\boldsymbol{u}(t)$ converges to the direction of maximal growth, $L(\boldsymbol{x})=L(\boldsymbol{x}, \boldsymbol{u})$ (with $\boldsymbol{u}(t)$ in this direction) satisfies (3.98). Thus $L(\boldsymbol{x})$ defines a local rate of divergence, which in average is equal to the maximal Lyapunov exponent λ_1.

3.4.2 *Contrast generated by fluctuations*

We can choose as state vector in (3.93) the set of concentrations

$$\begin{cases} \boldsymbol{c} = (c_k); \quad k = (j; \mathbf{x}) \quad (j = 1,\ldots, M); \\ c_k \geq 0; \quad \sum_k c_k = N, \end{cases} \tag{3.100}$$

of M chemical species in a volume described by V unit cells, where each cell is represented by its centre $\mathbf{x}$ in regular lattice. The number of dimensions in $\boldsymbol{c}$-space is thus $n = MV$. The total number of molecules has been fixed at N.

The equation of motion (3.93) now has to be generalized into

$$\dot{\boldsymbol{c}}(t) = \boldsymbol{F}(\boldsymbol{c}(t)) + \boldsymbol{w}(\boldsymbol{c}(t), t), \tag{3.101}$$

where $\boldsymbol{w}$ is a stochastic 'force,' characterized by

$$\begin{cases} < w(c, t) > = 0; \\ < w_k(c, t) \cdot w_{k'}(c, t') > = \eta c_k \left(\delta_{kk'} - \frac{c_{k'}}{N} \right) \delta(t - t'). \end{cases} \tag{3.102}$$

Here we have used the statistical fact that the fluctuations in squared number of a certain kind of molecules in a cell is proportional to the number of such molecules in that cell. The bracket on the right hand side of the second equation is necessary to keep the normalization, fixed in (3.100).

Since each orbit is a random walk depending on the stochastic variables w, we study an ensemble of such orbits described by a density function,

$$P(c, t) > 0; \quad \int d^n c \, P(c, t) = 1, \tag{3.103}$$

which then obeys a diffusion equation,

$$\dot{P}(c, t) = -\frac{\partial}{\partial c} \cdot (F(c) \, P(c, t)) + \eta \frac{\partial^2}{\partial c_k \, \partial c_{k'}} \left(c_k \left(\delta_{kk'} - \frac{c_{k'}}{N} \right) P(c, t) \right) \tag{3.104}$$

From this follows

$$\frac{d}{dt} < c > = \frac{d}{dt} \int d^n c \, P(c, t) \, c = < F(c) > \tag{3.105}$$

and

$$\frac{d}{dt} (< c^2 > - < c >^2) =$$

$$= 2(< c \cdot F(c) > - < c > \cdot < F(c) >) + 2\eta < \sum_k c_k \left(1 - \frac{c_k}{N} \right) >. \tag{3.106}$$

For $\eta = 0$, if P is originally a delta function, then it remains so indefinitely. With equilibrium concentrations $c_o = (c_{ko})$,

$$c_{ko} > 0; \quad \sum_k c_{ko} = N, \tag{3.107}$$

the contrast

$$C(\boldsymbol{c}) = \sum_k c_k \ln \frac{c_k}{c_{ko}} \tag{3.108}$$

shows the following behaviour:

$$\frac{\mathrm{d}}{\mathrm{d}t} \int \mathrm{d}^n \boldsymbol{c}\, P(\boldsymbol{c}, t)\, C(\boldsymbol{c}) =$$

$$= \int \mathrm{d}^n \boldsymbol{c}\, P(\boldsymbol{c}, t) \sum_k F_k(\boldsymbol{c}) \ln \frac{c_k}{c_{ko}} + N\, 2\eta MV \left(1 - \frac{1}{MV}\right), \tag{3.109}$$

where the first term is negative, since the deterministic process is dissipative (cf. (3.27)).

The second term in (3.109) is constant and positive, and represents a steady flow of contrast from microcosmos. In Fig. 3.1, this flow is indicated by upward arrows from the region of small r. The contrast flow per particle is $2\eta M$, if we neglect the correction due to normalization.

A stability analysis in the presence of fluctuations could be based on (3.104). One feature which may simplify such an analysis is that this equation is linear in $P(\boldsymbol{c}, t)$.

4. DISCRETE SYSTEMS AND COMPUTATION

There are numerous examples of systems that can be analysed as sequences of symbols, e.g., a written text, a DNA-sequence, or the input or output of a computer. The major part of this chapter applies to sequences of symbols, which may or may not carry a message. An application of the latter case is the sequence of zeroes and ones in a cellular automaton, to be discussed in Section 4.3. Cellular automata have the advantage that, although very simple, they may be rich enough to represent physical systems. Closely related to this is the fact that, for certain physical systems, the entropy is the thermodynamic average of a microscopic information-theoretical property, namely textual contrast.

Changes in symbol sequences can be viewed as actions by an abstract computer. Therefore, idealized computation and the physical conditions for it is a recurrent theme of this chapter.

The chapter ends with information theory applied to another type of discrete models, the input-output models of systems ecology and econometrics.

4.1 Information Theory of Symbol Sequences

Consider a sequence of symbols, taken from a finite set A with M elements. For simplicity, we assume the sequence to be doubly infinite. At each lattice site, the concentration or the probability for a symbol is either 0 or 1, which allows the formulas of Chapter 3 to be simplified.

If nothing is known a priori about frequencies and correlations within the sequence, the best binary coding gives an average code length of $\log M$ per symbol. (We use 'log' for the base two logarithm.) This length is the total information (in bits) per site of the sequence. We shall decompose the total information into chemical, correlational and textual contrast, and the correlational contrast can be further decomposed into contributions from different correlation lengths.

The a priori probability distribution is

$$c_i^{(0)} = \frac{1}{M}\,, \text{ for all symbols } i. \tag{4.1}$$

Denote the symbol at site x by $j(x)$. The probability $c_i(x)$ for the symbol i at position x can then be written

$$c_i(x) = \delta_{ij(x)}\,. \tag{4.2}$$

The correlational distribution $q^{(m)}$ of Section 3.3 is now equal to the distribution of m-length sequences and we write

$$p_m(i_1,\ldots,i_m) = \lim_{L\to\infty} \frac{1}{2L+1} \sum_{k=-L}^{L} \prod_{l=1}^{m} \delta_{i_l\, j(k+l)} \tag{4.3}$$

for the probability of the sequence $(i_1,\ldots,i_m)$ of symbols. Again, S_m denotes the entropy of m-point distributions

$$S_m = S[p_m] = -\sum_{i_1\ldots i_m} p_m(i_1\ldots i_m) \ln p_m(i_1\ldots i_m). \tag{4.4}$$

The total information k_{tot}, the contrast between the a priori distribution and $c_i(x)$, is

$$k_{\text{tot}} = K[c^{(0)};\, c(x)] = \ln M\,, \tag{4.5}$$

and can be decomposed into chemical contrast k_{chem}, correlational contrast k_{corr}, and (average) textual contrast $\bar{k}_{\text{tx}}$

$$k_{\text{tot}} = k_{\text{chem}} + k_{\text{corr}} + \bar{k}_{\text{tx}}\,. \tag{4.6}$$

Here, as in Section 3.3,

$$k_{\text{chem}} = K[c^{(0)};\, \bar{c}] = \ln M - S_1\,, \tag{4.7}$$

and (cf. (3.81-83))

$$k_{\text{corr}} = \sum_{m=2}^{\infty} k_{\text{corr}}^{(m)} = S_1 - \lim_{m \to \infty} \Delta S_m = S_1 - \Delta S_\infty \,, \tag{4.8}$$

$$\bar{k}_{\text{tx}} = \lim_{m \to \infty} \Delta S_m = \Delta S_\infty \,. \tag{4.9}$$

The average textual contrast (4.9) is identical to the *measure entropy*, s_μ, which is a widely used concept in dynamic systems theory, e.g., [Eckman and Ruelle 1985].

If correlations only up to length m are known, the m-length local textual contrast is

$$k_{\text{tx}}^{(m)}(x) = \ln \frac{p_{m-1}(j(x-m+1), \ldots, j(x-1))}{p_m(j(x-m+1), \ldots, j(x-1), j(x))} \,, \tag{4.10}$$

which gives the average m-length textual contrast

$$< k_{\text{tx}}^{(m)}(x) > = \Delta S_m \,. \tag{4.11}$$

This is of course larger than or equal to the saturated average

$$< k_{\text{tx}}^{(m)}(x) > \geq \bar{k}_{\text{tx}} \,, \tag{4.12}$$

but this does not necessarily hold for the local counterparts. Locally, an increase in correlation length may lead to an increased textual contrast.

The average textual contrast is the information per site that remains, when the information in correlations (including chemical contrast) is subtracted from the total contrast. In texts of natural languages $\bar{k}_{\text{tx}}$ is the information conveyed through the text, but in meaningless sequences $\bar{k}_{\text{tx}}$ is the random information used by the algorithm or process generating the sequence.

The decomposition of correlational contrast into contributions from different correlation lengths can be used to define a mean correlation length $\bar{m}$, where length is defined so that the distance between two adjacent cells is 1,

$$\bar{m} = \sum_{m=1}^{\infty} m \, \frac{k_{\text{corr}}^{(m+1)}}{k_{\text{corr}}} . \tag{4.13}$$

From this we can define a new quantity,

$$\eta = \bar{m} \, k_{\text{corr}} = \sum_{m=1}^{\infty} m \, k_{\text{corr}}^{(m+1)} = - \sum_{m=1}^{\infty} m \, \Delta^2 S_{m+1} , \tag{4.14}$$

which is identical to a measure for complexity (effective measure complexity) defined by Grassberger [1986a]. Simple crystals and random sequences have low complexity values, while systems with long correlations (discussed in Section 4.3) can achieve relatively high values. The Grassberger complexity can be written

$$\eta = \lim_{m \to \infty} \, (S_m - m \, \bar{k}_{\text{tx}}). \tag{4.15}$$

Thus η is a measure of the convergence of Shannon entropy S_m/m to average textual contrast as $m \to \infty$. Slow convergence rate means long correlation length $\bar{m}$ and a high value of η.

4.1.1 *Computer-generated texts*

To illustrate these concepts, we apply information theory to a written text. Shannon examplified the correlations between characters in English in this way.

When reading a text the message can be understood even if some words are misspelled. The presence of correlations in the text implies that a part of the information that is carried by a symbol is superflous or *redundant*, but this helps us to discover spelling errors. Consider an alphabet consisting of the 26 letters in English, a space, and a set of 5 punctuation marks, a total of 32 symbols. If we do not know anything about the frequencies or the correlations, each symbol carries 5 bits of information. We call this the total information per symbol. The total information can be separated into redundant information and textual information. The redundant information consists of chemical contrast, which is present if symbols occur with different frequencies, and correlational contrast, which can be decomposed into contributions from information in correlations of different lengths.

One can let a computer produce texts by giving it correlations from real texts [Hayes 1983]. In practice, a long text is made available to the computer program, which fetches the correlations of desired length from this *source* text. If the program uses 1-point correlations, i.e., the frequencies of single symbols from the source text, the generated text should reproduce the different symbols with the same frequencies as in the source text, but without correlations. If the program uses m-point correlations, and the sequence $j_1 j_2 \dots j_{n-1}$, with $n \geq m$, has been generated, the next symbol j_n is randomly chosen with the conditional probability for j_n, for given $j_{n-m+1} \dots j_{n-1}$, cf. (4.10),

$$P(j_{n-m+1},\dots,j_{n-1},j_n \,|\, j_{n-m+1},\dots,j_{n-1}) = \exp(-k_{tx}^{(m)}(n)). \qquad (4.16)$$

In the beginning of the simulation, for step $n < m$, the program is restricted to use correlations of length n.

In the following example we have used some 10 pages of a well-known book, mentioned in the reference list, as the source text. By increasing the correlation length we clearly see how more and more of the correlational information of the source text is reproduced. Already for 2-point correlations the language of the text can be guessed, and it is clearly identifiable for 3-point correlations.

1-point correlations:

Tdory d neAeeeko,hs wieadad ittid eIa c i lodhign un a a svmb i ee' kwrdmn.

2-point correlations:

Le hoin. whan theoaromies out thengachilathedrid be we frergied ate k y wee ' e the sle! se at te thenegeplid whe tly titou hinyougea g l fo nd

3-point correlations:

'Weed. Thed to dre you and a dennie. A le men eark yous, the sle nown ithe haved saindy. If - it to to it dre to gre. I wall much. 'Give th pal yould the it going, youldn't thave away, justove mouble so goink steace, 'If take we're do mennie.

4-point correlations:

I can light,' George tried in you and fire.' 'Nothen it and I want yourse, George some other ther. There's if his hand rolledad ther hisky, 'I little amonely we're we're with him the rain.

5-point correlations:

'I...I'm not running.'
The ranch, work on the time. Do you because you get somethings spready told you just him by heat to coloured rabbits. That's going grew it's like a whisky, place.

6-point correlations:

Million mice because it two men we'll sit by the future. We'll steal it. 'Aren't got it. 'About the fire slowly hand. 'I want, George,' he asked nervoulsly: "That's fine. Say it too hard forget other.

At the same time as the chemical and correlational contrasts are better reproduced, the value of the textual contrast also approaches the corresponding value of the source text. But in the source text the textual information carries a story or a message, while in the simulated text it is replaced by "random" information originating from the random number generator used by the program.

Brillouin [1956] estimated the textual contrast in English texts, and found it to be between 1 and 2 bits of information per symbol, with a total information of 4.7 bits. Thus the major part of the information is redundant.

4.2 Algorithmic Information Theory

The algorithmic information of an object is the size of the most compact description of it. Usually this is formalized for systems that can be described by sequences of zeroes and ones. Then the algorithmic information of a certain sequence is defined as the smallest program for a *general-purpose* computer to generate that sequence [Kolmogorov 1965, Chaitin 1966]. This leads to a formalism [Chaitin 1975] which is very close to information theory in the ordinary sense.

Two advantages with algorithmic information theory are that no probability measure is needed, and that the concepts can be applied to finite symbol sequences. A disadvantage is that the algorithmic information is usually not computable. This will be discussed in Section 4.2.3.

4.2.1 *The Turing machine*

A specific general-purpose computer, the universal Turing machine [Turing 1936-37, Hopcroft 1984], is chosen. This is a universal computer in the sense that it can simulate all other general-purpose computers.

The Turing machine consists of a control unit, a program tape, and a work tape, see Fig. 4.1. There is one tape head for reading from the program tape, and another one for reading from or writing on the work tape. The program is written in binary form on the program tape and it is executed by the control unit. From here on we shall not be concerned with the details of the program tape and its tape head.

The work tape, which is a doubly infinite sequence of symbols 0 and 1, contains both input and output. At each step the machine can do one of the following actions: Move the tape head left (L) or right (R), write a new symbol, 0 or 1, to the current tape cell, or halt. The control unit can be in different internal states, which, together with the program and the symbol in the current tape cell, determine the next action. If one allows for random influence in the choice of action, one would get a *non-deterministic* Turing machine.

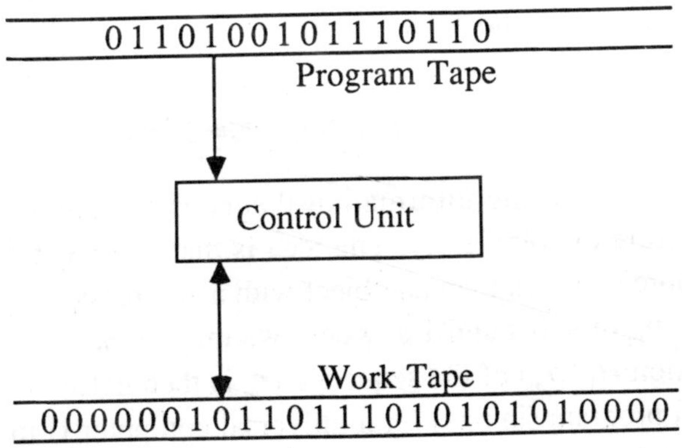

Fig 4.1 The Turing machine.

The set Q of states is finite and contains exactly one start state q_0 and a final state s. The set I of inputs is $\{0, 1\}$ and the set A of actions is $\{0, 1, L, R\}$. A program step is a mapping from $(Q \setminus \{s\}) \times I$ to $Q \times A$, and is written

$$q_j\, i\, a\, q_k \tag{4.17}$$

where $q_j \in Q \setminus \{s\}$ is the internal state, $i \in I$ is the symbol in the current tape cell, $a \in A$ is the action to be chosen, and $q_k \in Q$ is the new internal state. If $q_k = s$ the program halts. For example, the following program translates a finite sequence of 1's one step to the right, if the tape head is positioned somewhere in the sequence.

q_0 0 0 s
q_0 1 L q_1
q_1 1 L q_1
q_1 0 R q_2
q_2 1 0 q_3
q_3 1 R q_3
q_3 0 1 s

The first three lines move the tape head to the leftmost 1 in the sequence, the fourth line changes that to a 0, and the last two lines move the head to the right, change the first 0 to 1 and halts.

4.2.2 *Program length—complexity, computation time—depth*

The basic concept in algorithmic information theory is the algorithmic information as a measure of complexity. The idea is that an object that is hard to describe is more complex than an object with a shorter description. Consider a sequence α_m of m 0's and 1's, where we may let $m \to \infty$. The algorithmic information $H_U(\alpha_m)$ of the sequence α_m is then defined as the length of the shortest program (in bits) for a universal computer U to generate the sequence and then stop,

$$H_{\mathrm{U}}(\alpha_m) = \min_{\mathrm{U(P,\mathbf{0})}=\alpha_m} l(\mathrm{P}), \tag{4.18}$$

where $\mathbf{0}$ denotes an input (work tape) with only 0's, l(P) is the length of the program P, and U(P,$\mathbf{0}$) is the output (on the work tape) when $\mathbf{0}$ is used as input by the program P on the computer U. Let us choose one universal computer and drop the subscript U. The choice of universal computer leads to an ambiguity in an additive constant containing the information of how to simulate other universal computers,

$$H(\alpha_m) = H_{\mathrm{C}}(\alpha_m) + O(1), \tag{4.19}$$

where C denotes an arbitrary universal computer.

An m-length sequence α_m that cannot be described in a shorter way than 'print the sequence α_m', has the algorithmic information $H(\alpha_m) = m +$ $+ O(1)$ and is said to be 'algorithmically random'. However, a few of the m-length sequences β_m are ordered in some way and have a smaller algorithmic information $H(\beta_m) < m$.

The relative complexity of the sequence α given the sequence β is the length of the smallest program P generating α, starting with β on the work tape,

$$H_{\mathrm{U}}(\alpha \mid \beta) = \min_{\mathrm{U(P,}\beta)=\alpha} l(\mathrm{P}), \tag{4.20}$$

and as above

$$H(\alpha \mid \beta) = H_{\mathrm{C}}(\alpha \mid \beta) + O(1). \tag{4.21}$$

Obviously, $H(\alpha | \mathbf{0}) = H(\alpha)$, and $H(\alpha | \beta) \geq 0$ with equality only if $\alpha = \beta$. Another way to define relative complexity $H'(\alpha | \beta)$ is to put the minimal program β^* for β on the work tape [Chaitin 1975], $H'(\alpha | \beta) = H(\alpha | \beta^*)$.

Another property of a sequence, connected to its algorithmic information, is the time needed for a universal computer to generate the sequence.

There are numerous phenomena, such as chaotic behaviour, which have compact descriptions but take a very long time to compute or simulate. This time has been discussed by Bennett [1986] as a complement to the complexity and he calls it *depth.*

Given the smallest algorithm generating the sequence α_m, the depth $d_U(\alpha_m)$ is defined as the number of steps (or some other time unit) the machine U performs until it has generated the sequence and halted. In computer theory, problems are classified by the computation time as a function of the size n of the problem. The two most important classes contain problems that are solvable in polynomial and exponential time respectively.

The depth is a rather rough measure since a different choice of universal computer C leads to a depth $d_C(\alpha_m)$ which for large m has a logarithm that is proportional to the logarithm of $d_U(\alpha_m)$. The depth or the more general 'time complexity function' is a well studied concept in computer theory [Hopcroft and Ullman 1979].

4.2.3 *Relations to information-theoretical concepts*

It follows from Gödel's incompleteness theorem that the algorithmic information is usually not computable. However, under certain conditions, there is a close relation between textual contrast and algorithmic information [Zhvonkin and Levin 1970, Brudno 1977, Lindgren 1987]. Consequently, there is also a counterpart of correlational contrast in algorithmic information theory.

Consider an infinite sequence $\alpha = (i_1 i_2 ... i_m ...)$ which is the outcome of a stationary stochastic process and which has an average textual contrast $\bar{k}_{tx} > 0$. Let $h(\alpha_m)$ be the average algorithmic information per symbol of the sequence $\alpha_m = (i_1 i_2 ... i_m)$, $h(\alpha_m) = H(\alpha_m)/m$.

The correlational part, including the chemical contrast, of the total information is possible to describe, at least approximately, by a finite program, i.e., a finite amount of algorithmic information. However, since almost all sequences are algorithmically random [Martin-Löf 1966], the textual contrast must either be given explicitly, if the exact sequence is to be generated, or be available as the information in random numbers, if it is enough that the probabilities for all sequences are correctly generated. In both cases the amount of information needed is infinite with an average of $\bar{k}_{tx}$ per symbol. Thus, for almost all sequences with $\bar{k}_{tx} > 0$, we can conclude that, on the average, the algorihmic information is equal to the textual contrast

$$\lim_{m \to \infty} \frac{1}{m} H(\alpha_m) = h(\alpha) = \bar{k}_{tx}(\alpha). \tag{4.22}$$

Note that, in the limit $m \to \infty$, for almost all sequences α_m, there cannot exist any algorithm with a length that is shorter than $\bar{k}_{tx}(\alpha_m)m$ (see, e.g., [Leung-Yan-Cheong and Cover 1978]). Thus, to prove (4.22), it is sufficient to show how to construct an algorithm with length $H(\alpha_m)$ which generates the sequence α_m [Lindgren 1987]. The algorithm consists of one part which is a code for the sequence α_m, and another part which serves as a decoder. We divide the sequence α_m into n-length sequences ($n<<m$), where each n-length sequence is denoted by a new symbol γ, $\gamma \in \{\gamma_1,...,\gamma_N; N=2^n\}$, so that a new sequence $\Gamma_{m,n}$ of γ-symbols is formed. Given the symbol β, the probability for the next symbol γ_i is the conditional probability $p_\beta(\gamma_i) = p(\beta\gamma_i)/p(\beta)$, where $p(\beta\gamma_i)$ and $p(\beta)$ are the probabilities (4.3) for the sequences $\beta\gamma_i$ and β respectively. The code words for the symbols γ_i depend on the previous sequence β, and are chosen so that their lengths $l_\beta(\gamma_i)$ fulfill

$$-\log p_\beta(\gamma_i) \le l_\beta(\gamma_i) < -\log p_\beta(\gamma_i) + 1. \tag{4.23}$$

Since $\sum_i 2^{-l_\beta(\gamma_i)} \le 1$ we know from coding theory (e.g., [Ash 1965]) that there exists an 'instantaneous code' with the given lengths. In an instantaneous code no word is prefix of another one, so that the end of each word is given by the word itself. This is important since the coding depends on the previous symbol β. The average code word length l is

$$l = \sum_\beta p(\beta) \sum_{\gamma_i} p_\beta(\gamma_i)\, l_\beta(\gamma_i). \tag{4.24}$$

Taking the average of the logarithmic terms in (4.23) we get

$$-\sum_{\beta} p(\beta) \sum_{\gamma_i} p_\beta(\gamma_i) \log p_\beta(\gamma_i) = \sum_{\beta\gamma_i} p(\beta\gamma_i) \log \frac{p(\beta)}{p(\beta\gamma_i)} =$$

$$= S_2(\Gamma_{m,n}) - S_1(\Gamma_{m,n}) = \Delta S_2(\Gamma_{m,n}), \tag{4.25}$$

where S_1 and S_2 are entropies defined as in (4.4). To make the algorithm self-delimiting we add a prefix to each code word, which is one if the present symbol is the last one in the sequence $\Gamma_{m,n}$ and zero otherwise. Then the average code length per symbol in α_m is $l_{m,n}$=$(l$+1$)/n$, and the inequality (4.23) can be written

$$\frac{1}{n}\,(\Delta S_2(\Gamma_{m,n})+1) \le l_{m,n} < \frac{1}{n}\,(\Delta S_2(\Gamma_{m,n})+2). \tag{4.26}$$

Writing the entropies as functions of the original sequence α_m, we get

$$\frac{1}{n}\,\Delta S_2(\Gamma_{m,n}) = \frac{1}{n}\,(S_{2n}(\alpha_m) - S_n(\alpha_m)) \rightarrow \bar{k}_{\text{tx}}(\alpha_\infty), \tag{4.27}$$

if $n, m \rightarrow \infty$ as $n << m$. The length of the decoder grows exponentially with n and depends on the distributions $p_\beta(\gamma_i)$ but is independent of the length m of α_m. Thus, if we let $m, n, (\log m)/n \rightarrow \infty$, then $l_{m,n} \rightarrow h(\alpha)$ and $l_{m,n} \rightarrow \bar{k}_{\text{tx}}(\alpha)$, which give (4.22).

4.2.4 *Algorithmic redundancy and structure*

The difference between the length m of α_m and the algorithmic information $H(\alpha_m)$ can be interpreted as an algorithmic redundancy $R_{\text{A}}(\alpha_m)$,

$$R_{\text{A}}(\alpha_m) = m - H(\alpha_m). \tag{4.28}$$

Let $r_{\text{A}}(\alpha_m) = 1 - h(\alpha_m)$ denote its average per symbol. Then, according to (4.22), for the infinte sequence α, the algorithmic redundancy is equal to the information-theoretical redundancy $r(\alpha)$, i.e., the sum of the chemical and correlational contrast, $r_A(\alpha) = r(\alpha) \equiv k_{corr}(\alpha) + k_{chem}(\alpha)$.

Following Chaitin [1979] a measure of structure at different levels can be defined, using algorithmic information theory as follows. Let $\omega_d(\alpha_m)$ be a partition of α_m, consisting of non-overlapping sub-sequences of α_m with lengths $\leq d$, and let $L(\omega_d; \gamma)$ be the algorithmic information of the sub-sequence $\gamma \in \omega_d$ with length $l(\gamma)$. Further, let

$$L(\omega_d) = \sum_{\gamma \in \omega_d} \min(L(\omega_d; \gamma), l(\gamma)), \tag{4.29}$$

i.e., $L(\omega_d)$ is the sum of the algorithmic information of the sub-sequences, where a sub-sequence with no algorithm shorter than its length $l(\gamma)$, contributes with just $l(\gamma)$. (In other words, for those sub-sequences γ which do not have any shorter algorithm than 'print γ', the information in the print command is omitted.) Then minimize $L(\omega_d)$ over all possible partitions, and denote the minimum value $H_d(\alpha_m)$

$$H_d(\alpha_m) = \min_{\omega_d} L(\omega_d). \tag{4.30}$$

Thus, the difference $-\Delta_d H_d(\alpha_m) = H_{d-1}(\alpha_m) - H_d(\alpha_m)$ measures the algorithmic structures of length d.

For sufficiently small lengths $d \leq d_o$, the shortest algorithm that generates a d-length sequence β_d is 'print β_d'. Thus, $H_d(\alpha_m) = m$, for $d \leq d_o$. Note that $H(\alpha_m) = H_m(\alpha_m) + O(1)$, since the information in the 'print' command must be counted for $H(\alpha_m)$ if that would be needed in the shortest algorithm. These facts are used in the following decomposition of algorithmic redundancy

$$\sum_{d=2}^{m} -\Delta_d H_d(\alpha_m) = H_1(\alpha_m) - H_m(\alpha_m) =$$

$$= m - H(\alpha_m) + O(1) = R_A(\alpha_m) + O(1). \tag{4.31}$$

Here $-\Delta_d H_d(\alpha_m) \geq 0$ serves as the contribution to the algorithmic redundancy from algorithmic structures of length d, i.e., mutual algorithmic information which can be used to shorten the description of α_m. This corresponds to the contribution to the correlational contrast from the length d. However, there is no identity, not even for the infinite sequence α, since an algorithmic structure of length d need not be present as a correlation of length d, and vice versa. For $d \leq d_0$, $-\Delta_d H_d(\alpha_m) = 0$, even if the sequence α_m contains correlations of length d.

4.3 Cellular Automata

Cellular automata are abstract simulation models which are discrete in space and time, and the number of states in each lattice site or *cell* is finite. Originally they were introduced as idealized models of living systems, especially concerning the property of self-reproduction [von Neumann 1966]. The most famous example is perhaps the 'Life' mapping, which shows a very complex behaviour [Gardner 1971].

During the last twenty years, cellular automata have also been used to simulate a variety of physical systems (see, e.g., [Wolfram 1986b]), such as the microscopic motion in fluids [Salem and Wolfram 1986], the macroscopic concentrations in chemical systems [Winfree et al. 1985], the growth of crystals [Packard 1986], and abstract models for phase transitions [Kinzel 1985, Kaneko and Akutsu 1986].

Because of the high degree of simplicity and the possibility of complex behaviour, cellular automata can be studied in the search for evolutionary criteria for self-organizing systems. Investigation of complexity measures and related concepts for cellular automata have revealed interesting properties of their spatial organization and temporal behaviour [Wolfram 1983, 1984a, 1984b, 1986a, Grassberger 1986a, 1986b, Lindgren 1987].

4.3.1 *Definition of a cellular automaton*

In the following we will restrict ourself to one-dimensional, two-state cellular automata. Consider a doubly infinite sequence of cells i ($i \in Z$), in which each cell can be in either of two states, zero (0) or one (1). The state

s_i in cell i is changed from time t to time t+1, according to a rule which depends on the states in a finite neighbourhood of cell i,

$$s_i(t+1) = f(s_{i-r}(t), s_{i-r+1}(t), \ldots, s_i(t), \ldots, s_{i+r}(t)), \tag{4.32}$$

where r is the distance of influence from the surroundings. The case with r=1 is well studied and many of the 256 possible rules show complex behaviour. These rules are numbered according to Wolfram's notation

$$\mathrm{R} = \sum_{s_1, s_2, s_3 = 0}^{1} f(s_1, s_2, s_3)\, 2^{4s_1 + 2s_2 + s_3}. \tag{4.33}$$

In Fig. 4.2 the space-time pattern for rule 110 is shown. At time $t = 0$, the first line of the figure, the state in each cell is randomly chosen with equal probabilities for 0 and 1. The system is periodic, i.e., the first and the last (400th) cell are adjacent. As time goes on, correlations are built up and a periodic background pattern appears with moving structures overlayed.

4.3.2 *Information theory of cellular automata*

The entropy of a deterministic cellular automaton does not increase in time as it does for thermodynamical systems. The reason for this is that a cellular automaton can not generally be considered as closed in the thermodynamical sense, because the rules are not reversible. This leads to build-up of correlations, and a decrease in (average) textual contrast (Shannon entropy). (In this section we will deal only with *average* textual contrast, and, therefore, we drop the bar denoting average: $\bar{k}_{\text{tx}} \rightarrow k_{\text{tx}}$.)

The randomness, expressed by the textual contrast, is partially and irreversibly transformed into correlational or chemical contrast when complex structures evolve. The system can be regarded as closed or open with respect to random information (noise), giving deterministic or probabilistic rules. Only when the system is influenced by noise, can correlational information be transformed into textual information.

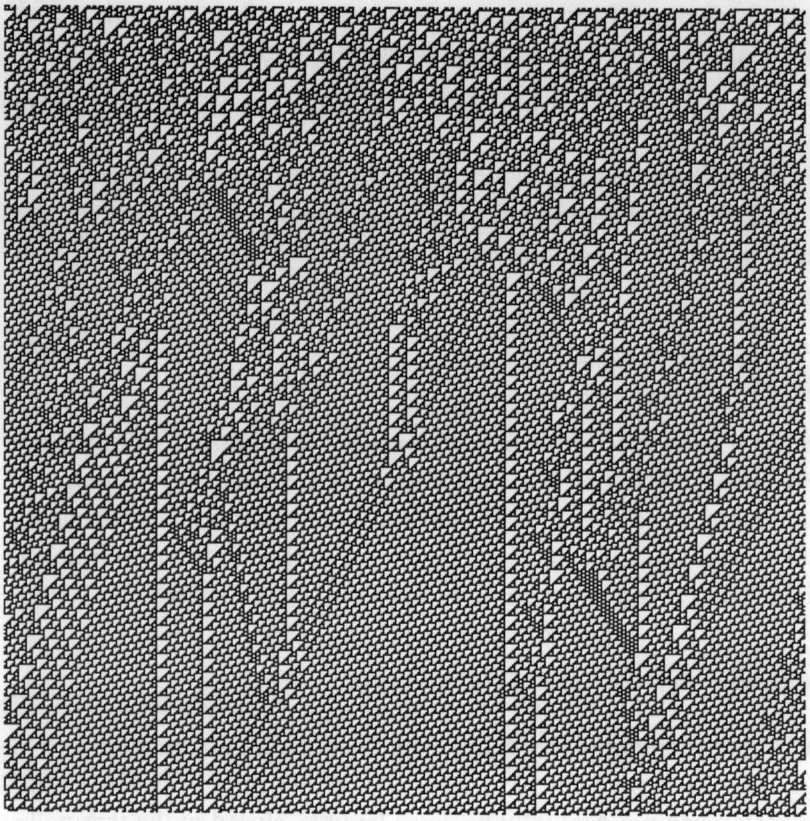

Fig. 4.2. The space-time pattern formed by rule 110.

If the cellular automaton rule is sufficiently reversible and is influenced by noise in an unbiased way, the textual contrast increases until the system is completely randomized.

The dynamic behaviour of one-dimensional cellular automata (4.32) can be studied with the concepts of Chapter 3 [Lindgren 1987]. The following treatment is easily extended to higher dimensions. We assume the initial state, a doubly infinite sequence $\Sigma(0)$, at $t = 0$ to be generated by a stationary stochastic process, giving spatially independent probabilities for all possible sequences. Consider a rule R that depends on neighbours up to the finite distance $r > 0$. At time t, the original sequence $\Sigma(0)$ has been transformed into the sequence $\Sigma(t) = (\ldots, s_{-1}(t), s_0(t), s_1(t), \ldots)$.

A sequence of $m+2r$ cells, with the joint state $\alpha_{m+2r}=(i_{1-r} \ldots i_{m+r})$, determines the joint state in a sequence of m cells, $\beta_m=\mathrm{R}_m(i_{1-r}\ldots i_{m+r})= =(j_1\ldots j_m)$. Denote the probability for an m-length sequence $\beta_m=(j_1\ldots j_m)$ at time t by $p_m(\beta_m;t)$,

$$p_m(\beta_m;\, t) = \lim_{L\to\infty} \frac{1}{2L+1} \sum_{k=-L}^{L} \prod_{i=1}^{m} ((s_{k+i} + j_i + 1) \bmod 2). \tag{4.34}$$

The ergodic theorem for stationary processes assures this limit to exist for almost all sequences Σ [Grimmett and Stirzaker 1982].

Deterministic rules

The probabilities for sequences at time t and t+1 are related through a $2^{m+2r} \times 2^m$ transfer matrix T_R for rule R, defined by

$$T_\mathrm{R}(\alpha_{m+2r}, \beta_m) = \delta(\mathrm{R}_m(\alpha_{m+2r}), \beta_m), \tag{4.35}$$

giving

$$p_m(\beta_m;\, t+1) = \sum_{\alpha_{m+2r}} T_\mathrm{R}(\alpha_{m+2r};\, \beta_m)\, p_{m+2r}(\alpha_{m+2r};\, t). \tag{4.36}$$

This leads to an entropy at time $t+1$

$$S_m(t+1) = -\sum_{\beta_m} p_m(\beta_m;\, t+1) \log p_m(\beta_m;\, t+1) \le S_{m+2r}(t). \tag{4.37}$$

The increase in textual contrast for one time step, $\Delta_t\, k_\mathrm{tx}(t)=k_\mathrm{tx}(t+1)-k_\mathrm{tx}(t)$, is then

$$\Delta_t k_\mathrm{tx}(t) = \lim_{m\to\infty} \left(\frac{1}{m}\, S_m(t+1) - \frac{1}{m+2r}\, S_{m+2r}(t)\right) =$$

$$= \lim_{m\to\infty} \left[\frac{1}{m+2r}\, (S_m(t+1) - S_{m+2r}(t)) + \left(\frac{1}{m} - \frac{1}{m+2r}\right) S_m(t+1)\right], \tag{4.38}$$

since, in the second line, the last term goes to zero in the limit and (4.37) holds. Thus we have shown that the textual contrast for a deterministic infinite cellular automaton cannot increase in time

$$\Delta_t k_{\rm tx}(t) \leq 0. \tag{4.39}$$

Almost reversible deterministic rules

Next we apply the same technique to a class of rules which can be regarded as reversible in an information-theoretical sense. Then we expect the textual contrast to be constant in time.

Consider a rule which is a one-to-one mapping of its last (or analogously its first) argument, given the values of the other ones. We write the rule in the form

$$\mathrm{R}(i_{-r},\ldots,i_r) = i_r + f(i_{-r},\ldots,i_{r-1}) \mod 2, \tag{4.40}$$

where f is a mapping from $\{0,1\}^{2r}$ to $\{0,1\}$. These rules are surjective, i.e., any sequence β_m has a predecessor α_{m+2r} such that $\mathrm{R}_m(\alpha_{m+2r})=\beta_m$. (If the rule R can be written as a sum over all its arguments, R is additive.) If the m-length sequence of states $\beta_m=(i_0 i_1 \ldots i_{m-1})$ at time t+1 is known, as well as the first $2r$ states of its predecessor $\alpha_{m+2r}=(j_{-r}j_{-r+1}\ldots j_0 j_1 \ldots j_{m-1+r})$ at time t, the whole sequence α_{m+2r} can be obtained by repeated use of the inverse of (4.40). Thus, knowing the state at time t+1, $2r$ bits of information are enough for achieving additional m bits of information about the state at time t. If we let $m \to \infty$, the average information per symbol that is needed to go one step back goes to zero. Since these rules are not reversible in the sense discussed in [Wolfram 1983], we call them *almost reversible.*

For each sequence β_m in (4.35), there are 2^{2r} sequences α_{m+2r} for which the delta-function side in (4.35) is equal to 1, if the rule is almost reversible. Then the transfer matrix has the properties

$$\sum_{\alpha_{m+2r}} T_{\mathrm{R}}(\alpha_{m+2r};\, \beta_m) = 2^{2r}\;; \tag{4.41}$$
$$\sum_{\beta_m} T_{\mathrm{R}}(\alpha_{m+2r};\, \beta_m) = 1,$$

which leads to a difference in entropy between time t+1 and t

$$S_m(t+1) - S_{m+2r}(t) = -\sum_{\beta_m} p_m(\beta_m;\, t+1) \log p_m(\beta_m;\, t+1) +$$
$$+ \sum_{\alpha_{m+2r}} p_{m+2r}(\alpha_{m+2r};\, t) \log p_{m+2r}(\alpha_{m+2r};\, t) \geq -2r\,. \tag{4.42}$$

Thus

$$\Delta_t k_{\mathrm{tx}}(t) \geq -\lim_{m\to\infty} \frac{2r}{m} = 0, \tag{4.43}$$

which together with the result (4.39) proves that infinite cellular automata governed by almost reversible rules conserve textual contrast

$$\Delta_t k_{\mathrm{tx}}(t) = 0. \tag{4.44}$$

Almost reversible rules influenced by noise

If we let the rule R be influenced by noise it should be possible for the textual contrast to increase. We assume that after the deterministic rule R has been applied in a cell, there is a probability q that the state in the cell will be changed. We denote such a probabilistic rule by the pair (R,q). Then the change in textual contrast in one time step can be divided into two parts, one coming from the deterministic rule R, $(\Delta_t k_{\mathrm{tx}})_{\mathrm{R}}$, given by (4.35), and another one due to the noise, $(\Delta_t\, k_{\mathrm{tx}})_{\mathrm{noise}}$. The system is open to random information and, as will be shown below, the influence of noise gives a non-negative contribution to the change in textual contrast. The influence of noise transforms the probability $p_m(\alpha_m)$ for the sequence α_m

$$\hat{p}_m(\alpha_m) = \sum_{\beta_m} T_{\text{err}}(\alpha_m; \beta_m)\, p_m(\beta_m). \tag{4.45}$$

The transfer matrix is

$$T_{\text{err}}(\alpha_m; \beta_m) = q^{H(\alpha_m, \beta_m)} (1-q)^{m - H(\alpha_m, \beta_m)}, \tag{4.46}$$

where $H(\alpha_m, \beta_m)$ is the Hamming distance between α_m and β_m, i.e., the number of positions in which the sequences differ. Thus the transfer matrix is symmetric and normalized,

$$\sum_{\beta_m} T_{\text{err}}(\alpha_m; \beta_m) = \sum_{\alpha_m} T_{\text{err}}(\alpha_m; \beta_m) = 1. \tag{4.47}$$

The entropy of the disturbed probability $\hat{p}_m$ is then

$$\begin{aligned} S[\hat{p}_m] &= -\sum_{\alpha_m} \hat{p}_m(\alpha_m) \log \hat{p}_m(\alpha_m) \geq \\ &\geq -\sum_{\beta_m} p_m(\beta_m) \log p_m(\beta_m) = S[p_m], \end{aligned} \tag{4.48}$$

where the inequality comes from (4.45) and (4.47) and the convexity of S. The entropy increase due to the noise can be written as a weighted sum of contrasts

$$\begin{aligned} (\Delta S_m)_{\text{noise}} &= S[\hat{p}_m] - S[p_m] = \\ &= \sum_{\alpha_m} \hat{p}_m(\alpha_m)\, K[T(\alpha_m,\cdot); T(\alpha_m,\cdot)\, p_m(\cdot)/\hat{p}_m(\alpha_m)] \geq 0. \end{aligned} \tag{4.49}$$

Thus, $(\Delta S_m)_{\text{noise}}$ vanishes only if $\hat{p}_m(\alpha_m) = p_m(\beta_m)$ for all α_m and β_m, that is for the totally uncorrelated distribution $\hat{p}_m = p_m = (1/2)^m$. The contribution to the increase in textual contrast is then

$$(\Delta_t k_{tx})_{noise} = \frac{1}{m}(\Delta S_m)_{noise} \, . \tag{4.50}$$

The textual contrast can increase or decrease depending on which is the bigger of $-(\Delta_t k_{tx})_R$ and $(\Delta_t k_{tx})_{noise}$. For almost reversible rules however, we know from (4.44) that $(\Delta_t k_{tx})_R=0$. Thus, for an infinite cellular automaton evolving according to the probabilistic rule (R,q), where R is an almost reversible rule and the error probability $q > 0$, the textual contrast increases in time until the sequence is completely randomized ($k_{tx}=1$)

$$\begin{cases} \Delta_t k_{tx}(t) > 0, & \text{if } k_{tx}(t) < 1; \\ \Delta_t k_{tx}(t) = 0, & \text{if } k_{tx}(t) = 1. \end{cases} \tag{4.51}$$

The results concerning the change in textual contrast in the time evolution for different classes of rules are summarized in Table 4.1.

Table 4.1. The change of textual contrast in time for different classes of rules.

	Deterministic	Probabilistic
Not *almost reversible*	$\Delta_t k_{tx}(t) \leq 0$	——
Almost reversible	$\Delta_t k_{tx}(t) = 0$	$\Delta_t k_{tx}(t) \geq 0$

4.3.3 *Simulation results*

In this section we will discuss three rules, which are examples of each of the three classes in Table 4.1. The rules are assumed to involve nearest neighbours only, $r = 1$.

As the first example we consider rule 110 (see Fig. 4.2), which is not almost reversible. According to (4.34), starting with a completely random sequence ($k_{tx} = 1$, $k_{corr} = 0$ and $k_{chem} = 0$), the time evolution leads to a decrease in the textual contrast until a stationary level is reached. Then some of the random information has been irreversibly transformed to chemical and correlational contrast. In Fig. 4.3 the growth of chemical and correlational contrast is shown as functions of time. The contributions from

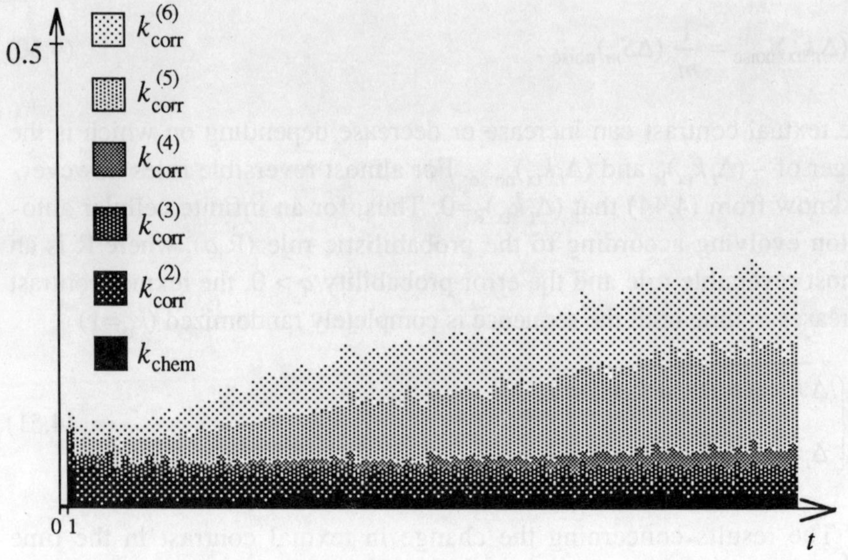

Fig. 4.3. The evolution of chemical and correlational contrast for rule 110. At each time, the chemical contrast and contributions to the correlational contrast from lengths $m \leq 6$ are calculated and drawn on top of each other. Initially, no correlations are present, but as the system evolves, the correlational contrast increases, especially due to the correlations of length 5 and 6. The system approaches a periodic state, in which the only contribution to the correlational contrast comes from lengths $m \leq 6$.

chemical contrast and correlational contrasts from lengths $m \leq 6$, see (3.83) and (4.7-8), are put on top of each other and are represented by black to light grey. Correlational contrasts of higher order are not calculated. The figure clearly shows how textual contrast is transformed to chemical and correlational contrast.

Next, we look at the time evolution of rule 195, see Fig 4.4, which is a very simple almost reversible rule: $s_i(t+1) = s_{i-1}(t) - s_i(t) + 1 \bmod 2$. We choose an initial state which is uncorrelated and has a density p $(0 < p < 1/2)$ of zeroes, giving $k_{chem} > 0$, $k_{corr} = 0$ and $k_{tx} < 1$. According to (4.44), this rule does not change the textual contrast. However, information may be transformed between chemical contrast and correlational contrast as well as between correlational contrasts of different lengths.

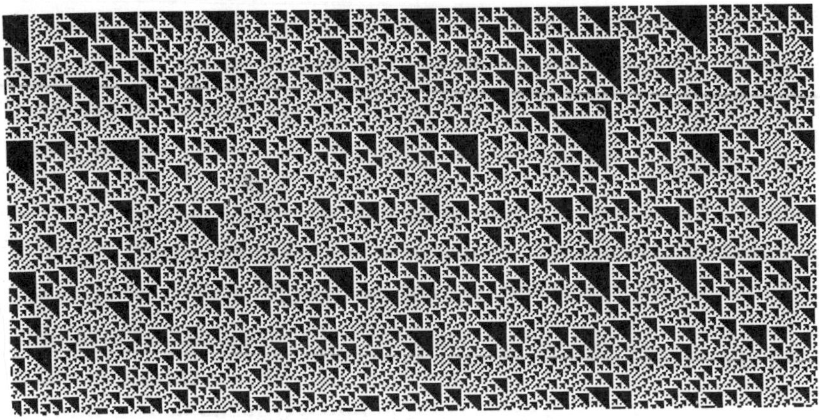

Fig. 4.4. The space-time pattern formed by the additive rule 195.

For the simulation of Fig 4.4, the density of zeroes $p = 0.1$ was chosen. As in the previous example, chemical and correlational contrasts up to length 6 were calculated, see Fig 4.5. Since the textual contrast is constant the figure shows that information flows from small to large correlation

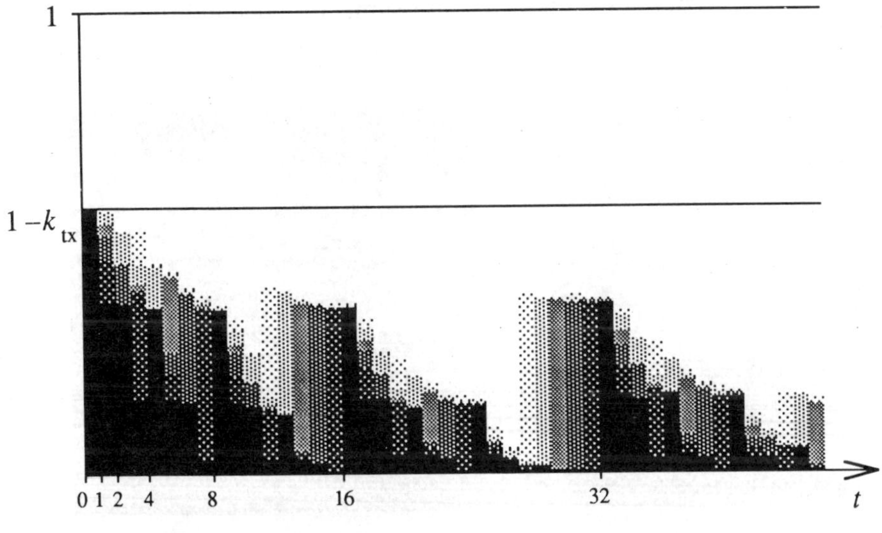

Fig. 4.5. The contribution to the correlational contrast from correlations of length $m \leq 6$ and the chemical contrast as functions of time for the additive rule 195, cf. Fig. 4.3.

lengths in the beginning, but after some time, some of the correlational contrast comes back and is transformed to chemical contrast again, and so the process continues. The time points t^* in which the chemical contrast takes its largest value are of the form $t^* = 2^n$ ($n = 1, 2, \ldots$). At these times there is no correlational information of lengths less than t^*, and the state can be regarded as a new uncorrelated initial state. This explains the self-similarity of the diagram, which cannot only be seen at these times, but also in corresponding times between two consecutive t^*. This self-similarity implies that, for almost all times as $t \to \infty$, there is neither chemical nor correlational contrast of finite order. Although the textual contrast is constant $k_{tx} < 1$, the system appears to be completely random. This process can be discussed in terms of Grassberger's complexity (4.14). At times t^*, the mean correlation length is proportional to t^*, leading to an ever increasing complexity for these times. We conjecture that this holds also for times $t \neq t^*$, but this remains to be proven.

In the third example we let the rule 195 be influenced by noise, see Fig 4.6. From (4.51) we know that all of the chemical and correlational contrast will be transformed into textual contrast, leading to a completely randomized system. By choosing the same initial conditions as in the previous example, and an error probability $q = 0.01$, we got the result

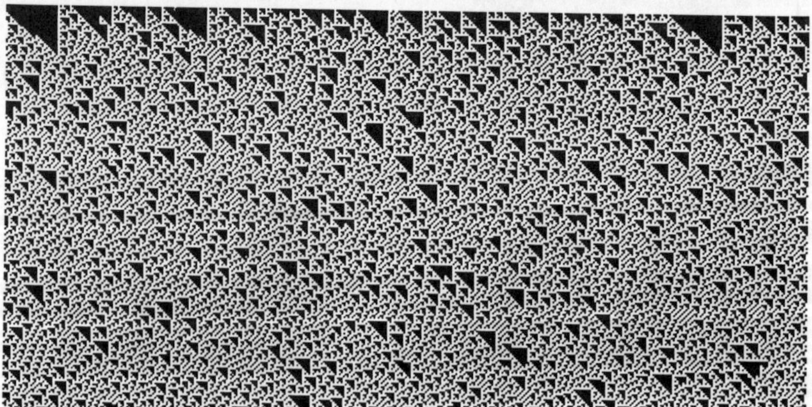

Fig. 4.6. The space-time pattern formed by the rule 195 when noise influences the evolution by changing the state with probability q=0.01 for each cell.

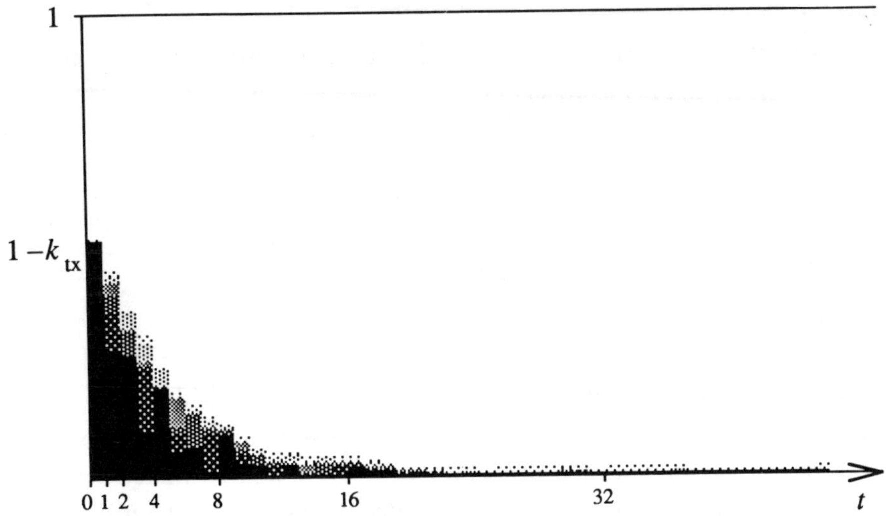

Fig. 4.7. The correlational contrasts from lengths $m \leq 6$, when rule 195 is influenced by noise.

shown in Fig 4.7. A comparison between this figure and Fig 4.5 clearly shows how sensitive the correlations are to random perturbations. The points of return of chemical contrast are almost completely gone.

For rules obeying the inequality (4.51), we thus have a "second law" telling us that the textual contrast (or, equivalently, the average Shannon entropy) increases in time. The last simulation shows an analogy to thermodynamical systems, where the entropy production can be viewed as a flow of information from the macroscopic level to correlations at the microscopic level, which are destroyed by stochastic fields (noise), as discussed in Section 3.2.

4.4 Thermodynamic Entropy and Textual Contrast

In this section we shall show that in a physical system, for which the thermodynamical limit exists, the ordinary (statistical mechanics) entropy is the thermodynamic average of the (microscopic) textual contrast of the system.

We consider a physical system with N available microsystems, and we assume it to be in the macrostate

$$P = \{p_i\}_{i=1}^{N} . \tag{4.52}$$

The entropy in binary units is then

$$S(P) = \sum_{i=1}^{N} p_i \log \frac{1}{p_i} = < -\log p_i >. \tag{4.53}$$

Thus it is an ensemble average over $-\log p_i$, a property which depends on the probability distribution P. Bennett has shown that if the macrostate is consicely describable, then the entropy S can be approximated with the ensemble average over the algorithmic information $H(i)$ of the microstates i. The following inequalities hold for the probability distribution P [Bennett 1982 (Section 6)]

$$S(P) \leq \sum_{i=1}^{N} p_i\, H(i) \leq S(P) + H(P) + O(1). \tag{4.54}$$

Here $H(P)$ denotes the minimum number of bits needed for describing the macrostate P, e.g., the equations of motion and boundary conditions which determine the macrostate. For a thermodynamical system, the entropy S is of the same order of magnitudes as the number of particles, say 10^{23} bits, while $H(P)$ corresponds to some thousand bits in a Monte Carlo program which generates the ensemble. Thus, the algorithmic information is a property of the microstate, and its average is equal to the thermodynamical entropy $S(P)$.

The relation (4.54) is conceptually important, but since $H(i)$ is not a computable function, we cannot directly use the relation as a tool for estimating the entropy. However, the equality between algorithmic entropy and textual contrast (4.22) then implies that the entropy $S(P)$ is an ensemble average of the textual contrasts $k_{tx}(i)$ of the microstates i. We shall show this equality directly without using information theory.

In the following we restrict ourselves to microstates that can be described as L-length sequences of zeroes and ones, $\alpha = (j_1, \dots, j_L)$. We denote the set of microstates by Ω. Then the macrostate P_L is a probability distribution over Ω,

$$P_{\mathrm{L}} = \{p_{\mathrm{L}}(\alpha)\}_{\alpha \in \Omega} \,. \tag{4.55}$$

We divide Ω into non-overlapping subsets Ω_k, such that each microstate in Ω_k has approximately the same correlations for lengths $m \leq M$. Here M is chosen to be large compared to typical correlation lengths that appear in the microstates, but small compared to L. (However there may occur low probability microstates with long correlation lengths for which such an inequality does not exist.)

Note that if a microstate α is generated by a non-deterministic algorithm (a Monte Carlo program) which uses relatively short correlation lengths $l_c << M << L$, then the information-theoretical concepts of Section 4.1 can be applied to that sequence. The microstates in Ω_k can then be regarded as generated by the same algorithm, but with different random numbers.

Let $Q_m(\alpha)$ denote the probability distribution over m-length sequences β in the microstate α,

$$Q_m(\alpha) = \{q_m(\alpha; \beta)\}_{\beta \in \mathrm{B}_m} \,, \tag{4.56}$$

where B_m denotes the set of all m-length binary sequences. Then the closeness between the elements in Ω_k can be expressed as

$$\frac{1}{m} K[Q_m(\alpha_1); Q_m(\alpha_2)] < \varepsilon << 1, \quad m \leq M, \quad \text{all } \alpha_1, \alpha_2 \in \Omega_k \,. \tag{4.57}$$

Let $p_\Omega(\Omega_k)$ be the probability in P_L for a randomly chosen microstate to belong to Ω_k, and let further $p_k(\alpha)$ be the probability for the microstate α, given that $\alpha \in \Omega_k$,

$$\begin{cases} p_\Omega(\Omega_k) = \sum_{\alpha \in \Omega_k} p(\alpha)\,; & (4.58) \\ p_k(\alpha) = \dfrac{p(\alpha)}{p_\Omega(\Omega_k)}\,, \quad \alpha \in \Omega_k\,. & (4.59) \end{cases}$$

A measure for how well Ω_k represents the correlational distributions $Q_M(\alpha)$, with $\alpha \in \Omega_k$, is given by

$$E_k^{(M)} = \frac{1}{L} \sum_{\alpha \in \Omega_k} p_k(\alpha) \log \frac{1}{p_k(\alpha)} - \sum_{\alpha \in \Omega_k} p_k(\alpha) < k_{tx}^{(M)}(\alpha) > =$$

$$= E_{k,1}^{(M)} + E_{k,2}^{(M)} + O(\varepsilon), \qquad (4.60)$$

with

$$\begin{cases} E_{k,1}^{(M)} = \dfrac{1}{L} \displaystyle\sum_{\alpha \in \Omega_k} p_k(\alpha) \log \frac{1}{p_k(\alpha)} - \frac{1}{M} \sum_{\beta \in B_m} q_M(\alpha_o;\beta) \log \frac{1}{q_M(\alpha_o;\beta)}\,; & (4.61) \\ E_{k,2}^{(M)} = \dfrac{1}{M} \displaystyle\sum_{\beta \in B_m} q_M(\alpha_o;\beta) \log \frac{1}{q_M(\alpha_o;\beta)} - < k_{tx}^{(M)}(\alpha_o) >, & (4.62) \end{cases}$$

where $\alpha_o \in \Omega_k$. The term $O(\varepsilon)$ must be added since a specific microstate α_o has been chosen in (4.61) and (4.62). Here $<k_{tx}{}^{(M)}(\alpha)>$ is the average M-length textual contrast (4.11) of α, for which the M-length distribution is $Q_M(\alpha)$.

The entropy $S(P_L)$ can be written

$$S(P_L) = \sum_k p_\Omega(\Omega_k) \sum_{\alpha \in \Omega_k} p_k(\alpha) \log \frac{1}{p_k(\alpha)} + \sum_k p_\Omega(\Omega_k) \log \frac{1}{p_\Omega(\Omega_k)} =$$

$$= \sum_k p_\Omega(\Omega_k)\, L \left(E_k^{(M)} + \sum_{\alpha \in \Omega_k} p_k(\alpha) < k_{tx}^{(M)}(\alpha) > \right) + S(P_\Omega) \qquad (4.63)$$

where $S(P_\Omega)$ is the 'partition entropy', defined by the last term in the first line. For the specific entropy $s(P_L) = S(P_L)/L$, we have

$$s(P_L) = \sum_{\alpha \in \Omega} p(\alpha) < k_{tx}^{(M)}(\alpha) > + \sum_k p_\Omega(\Omega_k)\, E_k^{(M)} + \frac{1}{L}\, S(P_\Omega)\,. \tag{4.64}$$

We assume that the correlations within almost all of the microstates stay finite as M, L and $L/M \to \infty$, i.e., the microstates can be regarded as being generated by non-deterministic algorithms using finite correlation lengths.

For fixed ε, the distance (4.57) between microstates in the subset Ω_k goes like $\varepsilon L^{-1/2}$, if Ω_k corresponds to a single algorithm, while the probabilities $p_\Omega(\Omega_k)$ for subsets Ω_k which do not correspond to any algorithm tend to zero. We assume that ε is sufficiently small, so that within each Ω_k for which $p_\Omega(\Omega_k)$ does not tend to zero, almost all states are generated by one single algorithm A_k. We also assume that the number of algorithms N_A which generate the microstates in P_L stays finite as $L \to \infty$. (This corresponds to the requirement in Bennett's theorem that the macrostate is concisely describable.)

The second term of $E_k^{(M)}$, (4.62), is of the order $O(\eta\, M^{-1})$, because of the definition of textual contrast (4.9-10) and its convergence property, expressed by the Grassberger complexity η (4.15). In the first term of $E_k^{(M)}$ (4.61), the entropy $S_M(Q_M(\alpha))/M$ differs from the entropy $S(P_A)/L$ of the probability distribution P_A on Ω, which is generated by the algorithm A_k, by a term of order $O(\eta\, M^{-1})$ (4.15). The first term of (4.61) can be viewed as an approximation of $S(P_A)/L$, in which low probability sequences falling outside Ω_k are not counted in the sum. The narrowing of the distribution P_A, expressed by the asymptotic decrease in distance by $\varepsilon L^{-1/2}$, then gives an additive term of the order $O(\varepsilon L^{-1/2})$. The choice of the specific α_0 gave an additive constant of the order $O(\varepsilon)$ in (4.60). Thus, the second term of $s(P_L)$ (4.64) is at most of the order $O(\varepsilon + \varepsilon L^{-1/2} + \eta M^{-1})$.

The average M-length textual contrast $<k_{tx}{}^{(M)}(\alpha)>$ approaches a saturation value $\overline{k}_{tx}(\alpha)$, as M increases, for almost all microstates. The partition entropy approaches a value less than $\log N_A$, and for fixed ε, the last term of the right-hand side of (4.64) goes like $1/L$. Note that ε can be arbitrarily small. Thus, in the thermodynamical limit $L \to \infty$, the specific entropy $s(P_L)$ for the macrostate is equal to the ensemble average of the average textual contrast $\overline{k}_{\mathrm{tx}}(\alpha)$ for the microstates α,

$$s(P_L) = \frac{1}{L} \sum_{\alpha \in \Omega} p(\alpha) \log \frac{1}{p(\alpha)} \to \sum_{\alpha \in \Omega} p(\alpha)\, \overline{k}_{\mathrm{tx}}(\alpha), \text{ as } L \to \infty. \tag{4.65}$$

The second law of thermodynamics tells us that the entropy $S(P)$ for a closed thermodynamical system increases in time. Equivalently, the ensemble average of the textual contrast for the microstates must increase in time. If the time evolution of P is governed by almost reversible microscopic rules (4.40) which are influenced by unbiased noise, the inequality (4.51) tells us that the average textual contrast of every microstate increases in time. Thus, the result (4.51) is a "second law" for the microstates which implies the second law of thermodynamics for probabilistic additive cellular automata.

4.5 Thermodynamics of Computation

We have seen (2.85) that exergy and information (in its physical sense) are simply different aspects of the same thing. For a given temperature T, one bit of information corresponds to an exergy quantity $kT \ln 2$.

The first and second laws of thermodynamics imply that ideally, i.e., in reversible processes, exergy is conserved, but in practice exergy is always spent. (The exception of a changing reference state is important cosmologically but for most practical applications, we have to deal with a fixed reference state.)

Thus the gain of one bit of information must be paid for by the loss of at least $kT \ln 2$ of exergy somewhere else. This was observed already by Szilard [1929] in a classical paper. The information needed by a Maxwell demon to discriminate between slow and fast molecules and then use them to drive a heat engine, would cost at least as much exergy as that obtained from the engine.

Landauer [1961] observed that logically irreversible steps in a computation necessarily involve dissipation. Thus, to throw away information is to throw away exergy. Thus, clearing a computer of garbage from intermediate computation steps by throwing away information involves a loss of exergy, i.e., an energy dissipation, of $kT \ln 2$ per bit.

In Fig. 4.8 we show a method to avoid throwing away garbage, i.e., avoiding exergy loss, using the fact that computation as well as copying, where the number of bits is unchanged, can in principle be done arbitrarily close to reversible. Copying can be done without production of garbage [Bennett 1982].

The way to minimize energy dissipation in computation is thus to work

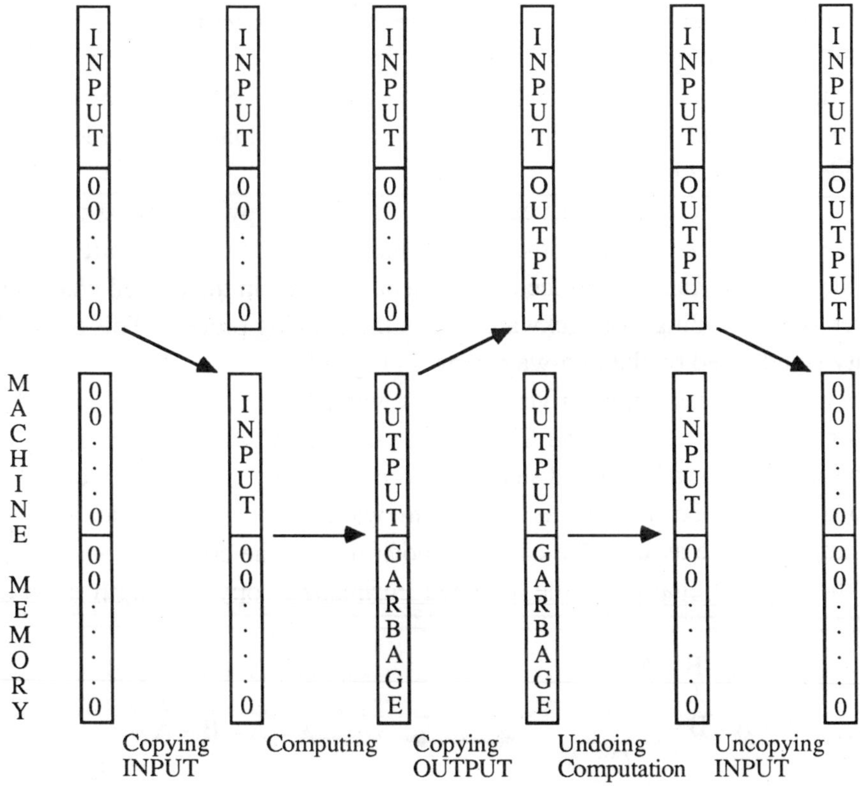

Fig. 4.8. Reversible computation .

with nearly reversible processes. A totally reversible process would not do because it could well undo itself. There must be something forcing the computation in one direction, making it to some degree irreversible and dissipative. This leads us to the 'Brownian computer'.

4.5.1 *Brownian computers*

As an example of a Brownian computer we will discuss a model of a chemical copying process, similar to the reproduction of DNA-sequences. More complex models of chemical computers are possible to construct, e.g., the chemical Turing machine [Bennett 1982].

Consider a macro-molecule which has the structure of an m-length chain of smaller molecules A and B. The molecule is copied step-by-step by building an identical chain attached to the original one, starting from one end. The process is going on in a solution, where A and B are connected to carrier molecules C. Assume that after n steps the situation in the left-hand side of Fig. 4.9 is reached.

If the molecule CA leaves A to the chain, the copying process proceeds one step further. On the other hand, since chemical reactions are reversible, a free C can bind to the last element (A or B) in the copied chain and remove it, leading to a backward step in the copying process. This possibility is indicated by the two-way arrow in Fig. 4.9.

The process can be driven forwards with a higher chemical potential μ for the molecules CA and CB than for the free C. But there is always a possibility that a backward step is performed instead of a forward one. The process can be viewed as a random walk with a drift velocity. For simplicity, we assume the chemical potential for the copying complex S_n (the original chain and the attached unfinished copy of length n) to be

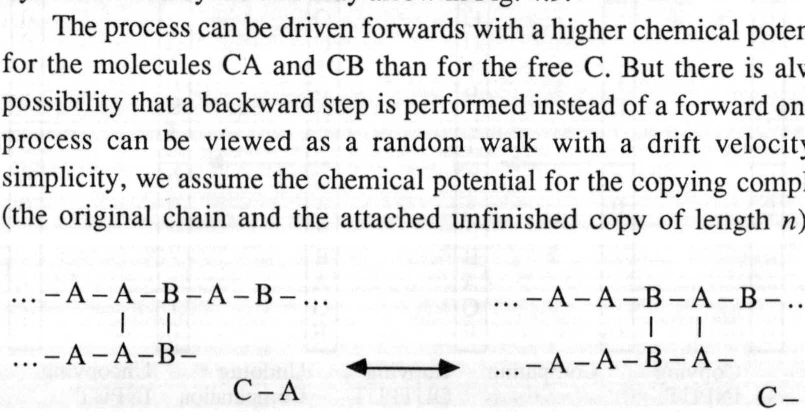

Fig. 4.9. A computational step in a copying process.

independent of step n, $\mu(S_n) - \mu(S_{n-1}) = 0$, and we put $\mu_{CA} = \mu_{CB}$. The chemical potentials are given by

$$\begin{cases} \mu_{CA} = kT \ln \dfrac{c_{CA}}{c_{CA}^{(o)}}; & (4.66) \\ \mu_{C} = kT \ln \dfrac{c_{C}}{c_{C}^{(o)}}, & (4.67) \end{cases}$$

where c_{CA} and c_C are the concentrations of CA and C respectively, and the superscript (o) denotes equilibrium.

The copying velocity, in the unit steps/second, is given by the reaction rate, $r = k_+ c_{CA} - k_- c_C$, where k_+ and k_- are rate constants. The average chemical entropy production σ is

$$\sigma = \frac{r}{T}\,(\mu_{CA} - \mu_C) = k\,(k_+ c_{CA} - k_- c_C) \ln \frac{k_+ c_{CA}}{k_- c_C}\,, \tag{4.68}$$

which vanishes in equilibrium. The rate of energy dissipation ε is the entropy production rate times the temperature, $\varepsilon = \sigma T$. The rate of entropy production per step in the copying process is

$$\sigma_1 = \frac{\sigma}{r} = k \ln \frac{k_+ c_{CA}}{k_- c_C}\,, \tag{4.69}$$

in which a temperature dependence can only be present in the rate constants.

In the process, the probability p for the next step to be in the forward direction is

$$p = \frac{k_+ c_{CA}}{k_+ c_{CA} + k_- c_C}\,, \tag{4.70}$$

and the ratio between the number of forward and backward steps is then

$$\frac{p}{1-p} = e^{-\sigma_1 / k}. \tag{4.71}$$

The dissipation rate can be made arbitrarily small by reducing the reaction rate, or equivalently by letting p approach 1/2. But this leads to a copying process with a higher total number of steps n_{step} per net step forward, $n_{step} = 1/(2p - 1), p > 1/2$. If there is an error probability γ associated with each step, a decreased dissipation rate will lead to an increased error probability Γ per net step forward. Note that an increase in temperature often leads to an increase in γ, although the forward/backward step ratio remains unchanged. For the total error probability we have

$$\Gamma = 1 - (1 - \gamma)^{n_{step}} , \tag{4.72}$$

and if we express the rate of entropy production per step as a function of the error probabilities γ and Γ we get

$$\sigma_1 = k \ln \frac{\ln(1 - \Gamma) + \ln(1 - \gamma)}{\ln(1 - \Gamma) - \ln(1 - \gamma)} . \tag{4.73}$$

For small Γ and low rate of entropy production, $\gamma << \Gamma << 1$ and $\sigma_1/k << 1$,

$$\frac{\sigma_1}{k} = 2 \frac{\gamma}{\Gamma} , \tag{4.74}$$

which shows the trade-off between error probability Γ and entropy production σ_1.

4.5.2 *Ballistic computation*

As a Gedanken experiment, consider a mechanical computer that uses billiard balls moving on a plane as signals, the 'ballistic computer'. One way to code the input is to have a number of channels parallel to the x-axis for each binary digit. Each channel consists of two half channels, one for 0 and one for 1, and has a ball in the relevant half-channel. All balls in the input enters the computer ($x = 0$) at the same time with the same speed v. The computer consists of fixed mirrors at an angle 45° to the axes. The balls may collide elastically with mirrors and with each other in such a way that they always move parallel to the x-axis or the y-axis. When the computation is finished all balls leave the computer ($x = L$) at the same time. In terms of Fig. 4.8 above, the copying/decopying may be considered as part of the computation process. This type of computer was proposed by Fredkin and Toffoli [1982].

In the ballistic computer, computation speed is proportional to the speed v of the balls. The kinetic energy of the input signal is carried over to the output and the kinetic energy tied to the 0's of the computer is not dissipated. Thus in principle a ballistic computer may be arbitrarily fast at zero dissipation.

The problem with the ballistic computer is the instability of ball orbits. Each collision increases the angular uncertainty by a factor $(1 + l/r)$, where l is the mean free path and r is the ball radius. Thus neglecting correlations, the individual error box grows by a factor $(1 + l/r)$ for each collision. This corresponds to the exergy loss per 'computational step' [Zurek 1984]

$$T\,\Delta S \cong kT \ln\left(1 + \frac{l}{r}\right) \cong kT \ln 2\,. \qquad (4.75)$$

This is thus the exergy that would have to be spent in order to stabilize the ball orbits sufficiently to make the computation work.

Partly this could be overcome by using instead of balls flat cylinders on small cars rolling on a square network of rails (cf. Bennett [1982]). The rails would provide the holonomic constraints needed to avoid the magnification factor $(1+l/r)$. The error volume would grow linearly instead of exponentially.

This shows that, classically, computation can in principle be done arbitrarily fast with arbitrarily small dissipation.

No system is totally free from dissipative forces such as friction. Therefore the Brownian computer is a more realistic model than the ballistic computer. As mentioned already, it is the Brownian type computers that are realized in DNA-replication and protein synthesis.

4.6 Stationary Flow Systems: Organizational Concepts

In a different context, that of system ecology or econometrics, the described system is composed of a finite number of discrete entities, 'compartments,' and 'flows' connecting them. The flows constitute an *input-output matrix* M [Leontief 1951]. It may describe energy (or exergy) flows or the flows of some other extensive quantity such as carbon (in ecology) or value (in econometrics).

If an input-output model were represented by a cellular automaton, the rules for change could be quite complicated. We shall deal here only with the simple case of a stationary system—or a periodic system averaged over one period. For this case, information-theoretical concepts have been introduced to measure the degree of organization or of evolution for the system. What we present here [Eriksson and Kåberger 1986] is a slight generalization of work by Ulanowicz [1981]. In Chapter 5 we shall discuss the applicability of these concepts to one model of prebiotic systems, that of a hypercycle.

Notation

We shall use *positive matrices*,

$$\begin{cases} \mathrm{A} = (A_{ij}) \quad (i = 1,\dots,m\,,\ \ j = 1,\dots, n); \\ A_{ij} \geq 0, \ \ \text{all } i, j\,; \\ A_{ij} > 0, \ \ \text{some } i, j\,. \end{cases} \tag{4.76}$$

An $(m \times n)$-matrix A is said to *restrict* another $(m \times n)$-matrix B,

$$\mathrm{A} \supset \mathrm{B}, \tag{4.77}$$

if $B_{ij} = 0$, whenever $A_{ij} = 0$, all i, j.

We shall introduce an $(m \times n)$ unit vector consisting only of 1's,

$$E_{ij} = E_{ij}^{(m,\,n)} = 1 \quad (i = 1,\dots, m\,,\ \ j = 1,\dots, n), \tag{4.78}$$

which may be used as a linear metric, defining the norm

$$\begin{cases} C = |\mathcal{M}|\,(S[\hat{\mathcal{M}}^{(o)}] - S[\hat{\mu}^{(o)}]) = 2\,|\mathrm{T}|\,S[\hat{\mathrm{T}}] = A + R\,; \\ A = |\mathcal{M}|\,(C[\hat{\mathcal{M}}] - C[\hat{\mu}]) = |\mathcal{M}|\,(K[\hat{\mathcal{M}}^{(o)};\,\hat{\mathcal{M}}] - K[\hat{\mu}^{(o)};\,\hat{\mu}]) = \\ \qquad\qquad\qquad\qquad = A_{\mathrm{M}} + A_{\mathrm{J}} + A_{\mathrm{L}} + A_{\mathrm{D}}\,; \\ R = |\mathcal{M}|\,(S[\hat{\mathcal{M}}] - S[\hat{\mu}]) = R_{\mathrm{M}} + R_{\mathrm{J}} + R_{\mathrm{L}} + R_{\mathrm{D}}\,, \end{cases} \tag{4.97}$$

where G was defined in (1.26), and

$$\begin{cases} A_{\mathrm{M}} = |\mathrm{M}|\,K[\hat{\mathrm{T}}\hat{\mathrm{T}}^{\sim};\,\hat{\mathrm{M}}]; \\ R_{\mathrm{M}} = |\mathrm{M}|\,S[\hat{\mathrm{M}}\,]. \end{cases} \tag{4.98}$$

Further

$$\begin{cases} A_{\mathrm{J}} = |\mathrm{J}|\,K[\hat{\mathrm{T}};\,\hat{\mathrm{J}}\,]; \\ A_{\mathrm{L}} = |\mathrm{L}|\,K[\hat{\mathrm{T}};\,\hat{\mathrm{L}}]; \\ A_{\mathrm{D}} = |\mathrm{D}|\,K[\hat{\mathrm{T}};\,\hat{\mathrm{D}}]; \end{cases} \qquad \begin{cases} R_{\mathrm{J}} = |\mathrm{J}|\,S[\hat{\mathrm{J}}\,]; \\ R_{\mathrm{L}} = |\mathrm{L}|\,S[\hat{\mathrm{L}}]; \\ R_{\mathrm{D}} = |\mathrm{D}|\,S[\hat{\mathrm{D}}], \end{cases} \tag{4.99}$$

and we define

$$\begin{cases} F_{\mathrm{J}} = A_{\mathrm{J}} + R_{\mathrm{J}} = \sum_{i=1}^{n} J_i \ln \hat{T}_i\,; \\ F_{\mathrm{L}} = A_{\mathrm{L}} + R_{\mathrm{L}} = \sum_{i=1}^{n} L_i \ln \hat{T}_i\,; \\ F_{\mathrm{D}} = A_{\mathrm{D}} + R_{\mathrm{D}} = \sum_{i=1}^{n} D_i \ln \hat{T}_i\,. \end{cases} \tag{4.100}$$

The quantities C, A, and R refer to the system with its external flows, i.e., the system considered as part of a larger system. The internal quantities A_{M}, and R_{M} are instead referring to the system itself (internal flows for each subsystem). In Ulanowicz's terminology F_{L} is a *tribute*, based on the *exports* L_i, paid by the system in order to be part of a larger system. Similarly, F_{J} is an *import tribute*, and F_{D} is a *dissipation fee* paid for irreversibility and non-zero speed. The upper limit for the internal ascendency A_M is reduced by the tributes and the dissipation fee, since

$$A_M + R_M = C - (F_J + F_L + F_D). \tag{4.101}$$

The ascendency A is a measure for the evolutionary state of a system functioning through flow exchanges. A may increase if the redundancy R decreases. A may also increase, for given import, export, and dissipation, if the internal flows M may increase. Like tributes and dissipation fee, redundancy is a price paid for the flexibility of having alternative pathways.

The quantities defined in (4.97) to (4.101) are extensive with respect to increase by a common factor of the flows within the system.

Moreover, the quantities defined here are chosen to be relatively robust with respect to the choice of subsystems. To see the effects of aggregation, let us use a double index for the throughput

$$\mathrm{T} = (T_{ia}) \tag{4.102}$$

and consider the aggregation related to

$$T_i^{(\mathrm{agg})} = \sum_a T_{ia}\,. \tag{4.103}$$

The entropy entering the expression for development capacity in C is

$$S[\hat{\mathrm{T}}] = \sum_{i,a} \frac{T_{ia}}{|\mathrm{T}|} \ln \frac{|\mathrm{T}|}{T_{ia}} = S[\hat{T}^{(\mathrm{agg})}] + \sum_i \hat{T}_i^{(\mathrm{agg})}\, S[\hat{t}^{(i)}]\,, \tag{4.104}$$

with

$$\hat{t}_a^{(i)} = \frac{T_{ia}}{T_i^{(\mathrm{agg})}}\,. \tag{4.105}$$

Thus, besides the entropy of the aggregated flow distribution, we also have to take the proper average of the internal entropies in the aggregated flows. Similar considerations hold for J, L, D, and M.

4.6.1 *Some simple model systems as examples*

In Table 4.2 [Eriksson and Kåberger 1986] we show some simple model systems, characterized by their Odum diagrams or the M-matrices [Odum 1971]. For each system A, R, and C are given. To facilitate comparisons all systems have their import J normalized to unity. Also all systems are closed in the sense that they have no export, $L = 0$. Hence their dissipation is also normalized to one, $D = 1$.

The entropy function for the interval $0 < x < 1$,

$$s(x) = s(1-x) = x \ln\frac{1}{x} + (1-x) \ln\frac{1}{1-x}, \tag{4.106}$$

occurs repeatedly in Table 4.2.

The simplest system in Table 4.2, system 1, has vanishing development capacity C, and hence vanishing A and R. System 2 has in general non-vanishing C, A, and R. The redundancy clearly comes from the fact that there are two dissipation flows. A increases as a function of x, i.e., A increases with increasing efficiency. In the limit of vanishing x, system 2 reduces to system 1.

System 3 is assumed to have two *producers* with no dissipation, and is thus a generalization of system 2 with maximum efficiency. It reduces to the system 2 with $x = 0$ in the limits $z = 0$ and $z = 1$. The parallel inflows and the parallel flows to the consumer produce redundancies $R_J = R_M = s(z)$. The ascendency is independent of how the inflow is shared between the producers.

System 4 has instead two *consumers* and one producer. For the case of no dissipation in the producer ($x = 1$), it is similar to system 3. (If arrows are reversed and dissipation and import are exchanged, so that producers and consumers exchange their roles, this is easily seen.) Again the ascendency is independent of the relative size of the parallel flows, thus A has the same value as in system 2. The increase in C as compared to system 2 is thus due to increased redundancy; R_M and R_D both increase by $x\, s(z)$. In the limit $x \to 0$, system 4 also collapses to system 1.

System 5 has two parallel systems of producer/consumer. The redun-

Table 4.2. Development capacity, ascendency, and redundancy for some simple models of ecological systems.

Odum diagram	$\mathcal{M} = \begin{pmatrix} 0 & 0 & 0 \\ D & M & L \\ 0 & J & 0 \end{pmatrix}$ $\begin{cases} \mathcal{M}^{(+)} = (0, T, \lvert J \rvert) \\ \mathcal{M}^{(-)} = (\lvert D \rvert, T, \lvert L \rvert) \end{cases}$	Development capacity, C Ascendency, A Redundancy, R Internal ascendency, A_M Internal redundancy, R_M
1) 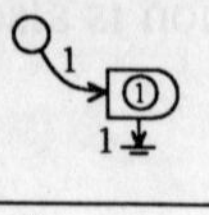	$\begin{pmatrix} 0 & 0 & 0 \\ 1 & 0 & 0 \\ 0 & 1 & 0 \end{pmatrix}$ $\begin{cases} \mathcal{M}^{(+)} = (0, 1, 1) \\ \mathcal{M}^{(-)} = (1, 1, 0) \end{cases}$	$\begin{cases} C = 0 \\ A = 0 \\ R = 0 \end{cases}$ $\begin{cases} A_M = 0 \\ R_M = 0 \end{cases}$
2) 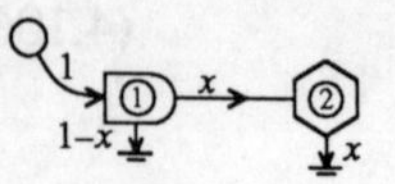	$\begin{pmatrix} 0 & 0 & 0 & 0 \\ 1-x & 0 & x & 0 \\ x & 0 & 0 & 0 \\ 0 & 1 & 0 & 0 \end{pmatrix}$ $\begin{cases} \mathcal{M}^{(+)} = (0, 1, x, 1) \\ \mathcal{M}^{(-)} = (1, 1, x, 0) \end{cases}$	$\begin{cases} C = 2((1+x)\ln(1+x) - x\ln x) \\ A = 4\left(\ln 2 - s\left(\frac{1+x}{2}\right)\right) + s(x) \\ R = s(x) \end{cases}$ $\begin{cases} A_M = x(2\ln(1+x) - \ln x) \\ R_M = 0 \end{cases}$
3) 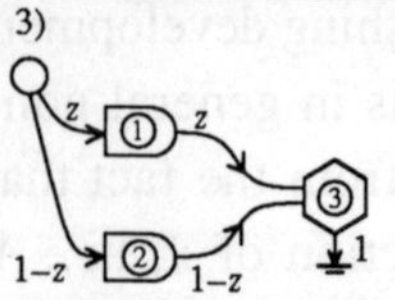	$\begin{pmatrix} 0 & 0 & 0 & 0 & 0 \\ 0 & 0 & 0 & z & 0 \\ 0 & 0 & 0 & 1-z & 0 \\ 1 & 0 & 0 & 0 & 0 \\ 0 & z & 1-z & 0 & 0 \end{pmatrix}$ $\begin{cases} \mathcal{M}^{(+)} = (0, z, 1-z, 1, 1) \\ \mathcal{M}^{(-)} = (1, z, 1-z, 1, 0) \end{cases}$	$\begin{cases} C = 2(2\ln 2 + s(z)) \\ A = 4\ln 2 \\ R = 2s(z) \end{cases}$ $\begin{cases} A_M = 2\ln 2 \\ R_M = s(z) \end{cases}$
4) 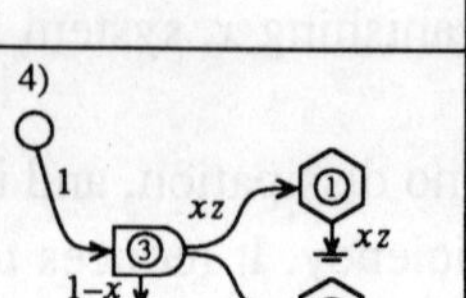	$\begin{pmatrix} 0 & 0 & 0 & 0 & 0 \\ xz & 0 & 0 & 0 & 0 \\ x(1-z) & 0 & 0 & 0 & 0 \\ 1-x & xz & x(1-z) & 0 & 0 \\ 0 & 0 & 0 & 1 & 0 \end{pmatrix}$ $\begin{cases} \mathcal{M}^{(+)} = (0, xz, x(1-z), 1, 1) \\ \mathcal{M}^{(-)} = (1, xz, x(1-z), 1, 0) \end{cases}$	$\begin{cases} C = 2((1+x)\ln(1+x) - x\ln x + x s(z)) \\ A = 4\left(\ln 2 - s\left(\frac{1+x}{2}\right)\right) + s(x) \\ R = s(x) + 2x s(z) \end{cases}$ $\begin{cases} A_M = x(2\ln(1+x) - \ln x) \\ R_M = x s(z) \end{cases}$
5) 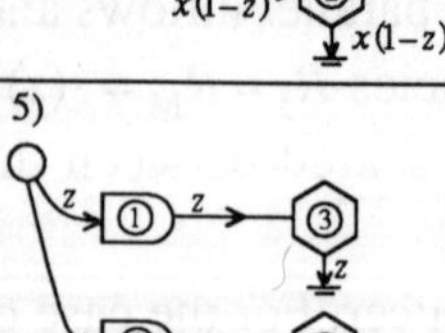	$\begin{pmatrix} 0 & 0 & 0 & 0 & 0 & 0 \\ 0 & 0 & 0 & z & 0 & 0 \\ 0 & 0 & 0 & 0 & 1-z & 0 \\ z & 0 & 0 & 0 & 0 & 0 \\ 1-z & 0 & 0 & 0 & 0 & 0 \\ 0 & z & 1-z & 0 & 0 & 0 \end{pmatrix}$ $\begin{cases} \mathcal{M}^{(+)} = (0, z, 1-z, z, 1-z, 1) \\ \mathcal{M}^{(-)} = (1, z, 1-z, z, 1-z, 0) \end{cases}$	$\begin{cases} C = 4(\ln 2 + s(z)) \\ A = 4\ln 2 + s(z) \\ R = 3s(z) \end{cases}$ $\begin{cases} A_M = 2\ln 2 + s(z) \\ R_M = s(z) \end{cases}$
6) 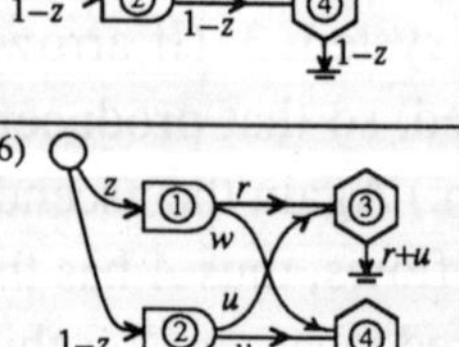 $\begin{cases} r = zx \\ w = z(1-x) \end{cases}$ $\begin{cases} u = (1-z)y \\ v = (1-z)(1-y) \end{cases}$	$\begin{pmatrix} 0 & 0 & 0 & 0 & 0 & 0 \\ 0 & 0 & 0 & r & w & 0 \\ 0 & 0 & 0 & u & v & 0 \\ r+u & 0 & 0 & 0 & 0 & 0 \\ w+v & 0 & 0 & 0 & 0 & 0 \\ 0 & z & 1-z & 0 & 0 & 0 \end{pmatrix}$ $\begin{cases} \mathcal{M}^{(+)} = (0, z, 1-z, r+u, w+v, 1) \\ \mathcal{M}^{(-)} = (1, z, 1-z, r+u, w+v, 0) \end{cases}$	$\begin{cases} C = 2(2\ln 2 + s(z) + s(zx + (1-z)y)) \\ A = 4\ln 2 + s(zx + (1-z)y) - zs(x) - (1-z)s(y) \\ R = 2s(z) + s(zx + (1-z)y) + zs(x) + (1-z)s(y) \end{cases}$ $\begin{cases} A_M = 2\ln 2 + s(zx + (1-z)y) - zs(x) - (1-z)s(y) \\ R_M = s(z) + zs(x) + (1-z)s(y) \end{cases}$

dancy has three equal contributions, $R_J = R_M = R_D = s(z)$. In this case A depends on how the inflow is divided between the two pathways, and A is maximal for an equal distribution. The same holds of course for R and hence also for C.

System 6 also contains two producers and two consumers. Now also crossed flows are included. In this case A has its maximal value $A = 4 \ln 2$ for $x = y$, independently of z. In the limits $x = y = 1$ (or $x = y = 0$), it reduces to system 3. In the limit $x = 1, y = 0$ (or $x = 0, y = 1$) it reduces to system 5. In the limits $z = 0$ or $z = 1$ system 6 reduces to system 4 with $x = 1$.

5. STRUCTURE IN CHEMICAL SYSTEMS

Beginning as late as in the mid-sixties, there has been an explosive development of both experimental and theoretical research on chemical systems evolving far from equilibrium (see, e.g., Glansdorff and Prigogine [1971], Nicolis and Prigogine [1977], Blumenfeld [1981], Vidal and Pacault [1981] and [1984], Field and Burger [1984]). We shall here discuss only systems in which structure emerges through chemical reactions, not because other systems are less interesting, but rather because of lack of space for a comprehensive treatment. Exotic behaviour such as multistability, oscillations, spatial self-organization and chaotic evolution, has been studied theoretically and found experimentally in numerous chemical systems. The interest has not only focused on these phenomena as such, it has also been turned towards more general and fundamental issues, such as the thermodynamics of nonlinear, far-from-equilibrium systems, and possible implications for biological systems. Biological or living systems are (among other things) polyfunctional well-ordered chemical systems. They use inputs from the environment to perpetuate themselves. Characteristic features are the spontaneous creation, maintenance, and reproduction of order.

The behavior of these systems is often so complicated that it is impossible to study it in full detail. The usual procedure is instead to construct a simple representative mathematical model with one or a few characteristic features, and to analyze the behavior of this model under various circumstances. It is customary to make the following simplifying assumptions: the system is assumed to be *isothermal*; the *local equilibrium approximation* discussed in Section 2.1.2 is assumed to be valid; if the reactions and the reaction laws are explicitly specified, they obey the *law of mass action* in ideal gases.

Most self-organizing chemical systems can conveniently be sorted, according to the kind of structure they have, into one of three categories: the structure can be *spatial*, *temporal*, or *spatio-temporal* (see, e.g., Vidal and Pacault [1981, 1984], Gurel and Gurel [1983], Jäger and Murray [1984], Rensing and Jaeger [1985]).

A spatial structure is a stationary (time-independent) geometrical structure. Some models of such structures were discussed in Chapter 3. Experimentally, the best-known examples are the Liesegang rings, and various interface patterns. Turing [1952] made an early theoretical analysis of biochemical structures. We note that, for obvious reasons, a long-lived stationary geometrical structure in a chemical system is likely to be two-dimensional.

A spatio-temporal strucure is a moving geometrical pattern, e.g., a concentration wave. The famous Belousov-Zhabotinsky-reaction is an example of an experimental system which exhibits this kind of structure [Tyson 1976]. Numerous mathematical models with such behaviour have been constructed, e.g., the Brusselator model [Nicolis and Prigogine 1977].

A temporal structure, a 'chemical oscillator,' is a homogeneous system with at least one time-dependent concentration [Rensing and Jaeger 1985]. Again, the Brusselator model and the Belousov- Zhabotinsky-reaction can serve as examples. We note that the homogeneity assumption implies that the mixing processes (diffusion and stirring) are fast compared to the chemical reactions.

5.1 Chemical Temporal Structures

Here, we shall discuss some thermodynamic aspects of chemical oscillators, e.g., the entropy production, and also the flow of information. The inherent dissipation of such systems, i.e., their entropy production, can be viewed as a cost which has to be paid by the systems. It is then natural to investigate, as will be done below, under what conditions a minimization of this cost turns out to be 'favorable' to the system. Another way of putting this is to ask whether a minimum entropy production principle serves as an evolution criterion for a particular kind of systems. Clearly, in a living organism such a selection criterion decreases the need for steady compensation of entropy production, as pointed out by Schrödinger [1944], and thus decreases the subsistence exergy input, which may be an essential cost for such systems.

5.1.1 *Entropy production and evolution criteria*

Dissipative structures are different from many of the systems studied in thermodynamics, in that they are *open* systems; energy and material are passing through them and are used by them. In view of the second law for closed systems, this openness is a necessity, since, during any small time interval, the change in entropy of any system, open to exchange of matter and energy from its surroundings, is

$$\Delta S = \Delta S^{(i)} + \Delta S^{(e)}. \tag{5.1}$$

Here $\Delta S^{(e)}$ is the change in the entropy of the system due to exchanges with the surroundings, and $\Delta S^{(i)}$ is the change resulting from irreversible processes ocurring inside the system. The second law states that $\Delta S^{(i)} \geq 0$ always, with equality only when no irreversible processes occur. Since $\Delta S^{(e)}$ is zero for an isolated system it follows that a system with $\Delta S<0$ *must* be an open system; only an open evolving system may carry out a process where its entropy is smaller in the end than in the beginning. The open system is maintained in its ordered state by exporting entropy to its surroundings, as heat and degraded chemical substances.

In the discussion above, we have used the entropy S, which is usually defined through equilibrium thermodynamics. There is therefore no guarantee that the same definition can be extended to non-equilibrium systems. The customary way to get around this is to assume local thermodynamic equilibrium (cf. Secion 2.1), so that a macroscopic evaluation of entropy production and entropy flow is possible also for open, non-equilibrium systems. Such an evaluation is based on a definition of a local entropy density which fulfils the local form of the Gibbs relation (2.38).

We showed in Section 2.1 that the entropy production density σ can be written as a sum of products of 'generalized flows' J_i and 'generalized forces' X_i [Glansdorff and Prigogine 1971] (cf. (2.45)). Generally, the relations between forces and flows are non-linear. Both forces and flows vanish in equilibrium. Near equilibrium, where the forces are small, the flows can be represented by the first non-vanishing term in a power series expansion:

$$J_i = \sum_j L_{ij} X_j \,, \qquad \text{where } L_{ij} = \left(\frac{\partial J_i}{\partial X_j} \right)_{\text{eq}} . \tag{5.2}$$

The field of 'linear non-equilibrium thermodynamics' deals with systems where this approximation is applicable. Within this field, a number of evolution criteria have been found, e.g., showing the relations between such thermodynamic quantities as the entropy production and the stability properties of stationary states. One such evolution criterion, the '*minimum entropy production principle*,' was derived in Section 2.1.3. It states that when the Onsager relations hold, the entropy production in a close-to-equilibrium stationary state has a minimum value compatible with the prescribed conditions. Furthermore, since

$$\frac{\mathrm{d}\sigma}{\mathrm{d}t} < 0 \tag{5.3}$$

in a neighborhood to the stationary state, σ can be used as a Lyapunov function, which guarantees the local stability of the stationary state. For this reason, the role of the entropy production in these systems is somewhat like that of the thermodynamic potentials in equilibrium theory.

In a chemical system where the heat conduction, diffusion and internal friction terms in (2.45) can be neglected, the local rate of entropy production per unit volume, σ, can be written

$$\sigma = -\sum_k A_k w_k \,, \tag{5.4}$$

where w_k is the reaction rate per unit volume in reaction k, and where the affinity A_k is given by (2.42),

$$A_k = -\sum_i \mu_i \nu_{ik} \,. \tag{5.5}$$

In the following, the chemical potential is taken to be given by an ideal-gas expression (cf. (2.77)) of the form

$$\mu_i = \mu_i^\circ + RT \ln x_i \,, \tag{5.6}$$

where μ_i° is a standard state chemical potential and x_i denotes the molar fraction of i. The reaction rates are assumed to be given by mass-action kinetics.

The linearity assumption is quite restrictive for chemical reactions, it amounts to the assumption $A_i << kT$. This is often unrealistic, particularly in living information-storing systems, since it implies a strong influence of thermal noise which will cause low reproduction accuracy, as we shall discuss later.

When the relations between forces and flows are non-linear the evolution criteria found so far (see, e.g., Nicolis and Prigogine [1977]) are less general.

5.1.2 *Entropy production in oscillating systems*

For an oscillating system, the instantaneous rate of entropy production is not the significant entity, rather it must be the average entropy production during a motion returning the system to its initial state. If a minimum entropy production principle is to be a selection criterion for these systems, it should therefore be based on the time-average of the entropy production. The natural question is then whether the average entropy production is lower on a periodic trajectory than in a stationary state.

We shall therefore here study the rate of dissipative losses, i.e., the rate of entropy production, due to chemical reactions in autonomous chemical oscillators. The quantity of interest is the difference between the average entropy production rate in a stable periodic oscillatory mode and in the corresponding unstable stationary state. This difference is one relevant measure of the efficiency of oscillatory behaviour, with possible implications for, e.g., metabolic processes.

We may note in passing that the entropy of mixing associated with the flows of reactants and products in a theoretical reaction system can be of arbitrary sign and magnitude.

Let the vector $\boldsymbol{x}(t)$ denote the state of a general system obeying the equations of motion

$$\dot{\boldsymbol{x}} = \boldsymbol{F}(\boldsymbol{x}(t);\alpha) , \tag{5.7}$$

where α is a scalar control parameter characterizing $\boldsymbol{F}$. Let $\boldsymbol{x}^*(\alpha)$ be a stationary state of the system (5.7), i.e.,

$$\boldsymbol{F}(\boldsymbol{x}^*) = 0. \tag{5.8}$$

Assume that there exists a periodic trajectory T with period τ for a certain parameter value α_0, i.e., for $\boldsymbol{x}(t) \in \mathrm{T}$,

$$\boldsymbol{x}(t) = \boldsymbol{x}\,(t + \tau). \tag{5.9}$$

Then, using the expression (5.10) for the entropy production rate the average entropy production rate is

$$\sigma_{\mathrm{T}}(\alpha_o) = \frac{1}{\tau} \int_0^{\tau} \sigma(\boldsymbol{x}\,(t))\,\mathrm{d}t\,, \tag{5.10}$$

where $\boldsymbol{x}(t) \in \mathrm{T}$. As indicated, it can be treated as a function of α_0. The difference in the rate of entropy production of a stable periodic trajectory and in an unstable steady state is then

$$\Psi(\alpha) = \sigma_{\mathrm{T}}(\alpha_o) - \sigma(\boldsymbol{x}^*(\alpha_o)). \tag{5.11}$$

(A similar variable can be introduced for parameter values such that there exist both stable and unstable closed trajectories.)

Generally, the character and local stability properties of the stationary states are evaluated via the eigenvalues, λ_i, of the Jacobian matrix of $\boldsymbol{F}$(cf. Section 3.4.1), evaluated at the stationary state. The eigenvalues can be used to classify the stationary state, e.g., 'focus' means that Im(λ) is non-zero for two complex-conjugate eigenvalues λ, 'node' indicates that all eigenvalues are real. A stationary state is stable if Re(λ)<0; it is unstable, and thus oscillations may be possible, if Re(λ) >0.

Let α_H be the *Hopf bifurcation* parameter value, i.e., for α close to α_H the Jacobian matrix (evaluated at $\boldsymbol{x}^*(\alpha_H)$) has two complex conjugate eigenvalues which can be written

$$\lambda_k = \varepsilon(\alpha) \pm i\omega(\alpha), \qquad k = 1, 2, \tag{5.12}$$

and where, by definition, $\varepsilon(\alpha_H) = 0$, $\omega(\alpha_H) \neq 0$. Except in pathological cases there exist two corresponding orthogonal eigenvectors which are also orthogonal to all other eigenvectors. These two eigenvectors define a plane in state space, and using them, two orthogonal unit vectors, say $\boldsymbol{e}_1$ and $\boldsymbol{e}_2$, in this plane, can be defined so that for sufficiently small positive $\varepsilon(\alpha)$ (i.e., α close to α_H) the closed trajectory is given by the circle

$$\boldsymbol{x}(t) = \boldsymbol{x}^*(\alpha_H) + r(\alpha)\,(\boldsymbol{e}_1 \cos\omega(\alpha)t + \boldsymbol{e}_2 \sin\omega(\alpha)t)\,, \tag{5.13}$$

where $r(\alpha_H)$=0.

Expansion of Ψ around the stationary state yields

$$\Psi = \frac{1}{\tau}\int_0^\tau \left\{ (\boldsymbol{x}(t) - \boldsymbol{x}^*(\alpha_H)) \left(\frac{\partial\sigma}{\partial \boldsymbol{x}} (\boldsymbol{x} = \boldsymbol{x}^*(\alpha_H)) \right) + \right.$$

$$\left. + \frac{1}{2} (\boldsymbol{x}(t) - \boldsymbol{x}^*(\alpha_H)) \cdot (\Xi(\boldsymbol{x}(t) - \boldsymbol{x}^*(\alpha_H))) + \ldots \right\} \mathrm{d}t, \tag{5.14}$$

where the matrix Ξ is the Hessian matrix evaluated at $\boldsymbol{x}^*(\alpha_H)$,

$$\Xi = \left\{ \frac{\partial^2\sigma}{\partial x_i\, \partial x_j} \right\}_{\boldsymbol{x}=\boldsymbol{x}^*(\alpha_H)} \tag{5.15}$$

The integral over the first term in (5.14) vanishes when the expression (5.13) is inserted. Thus, to second order,

$$\Psi = (e_1\cdot(\Xi e_1) + e_2\cdot(\Xi e_2))\ \frac{r^2}{4}\ . \tag{5.16}$$

This means that the sign of Ψ in a neighborhood of α_H is determined by this sum of quadratic forms. If Ξ is positive or negative definite for α close to α_H then this obviously determines the sign of Ψ in this neighborhood. This is of course no guarantee that Ψ will be always positive or negative.

There seems to be no other rule for determining whether Ψ is negative or positive at a Hopf bifurcation. The sign of Ψ may, in general, be positive or negative. In fact, the example below shows that both cases occur.

Example: a model with periodic oscillations

The behaviour of Ψ will be illustrated by the following model system of chemical reactions

$$\begin{aligned} A+Z &\rightleftarrows X+Z \\ Z &\rightleftarrows 2X \\ X+Y &\rightleftarrows B \\ 2Z &\rightleftarrows Y+C \\ Y &\rightleftarrows D \end{aligned} \tag{5.17}$$

It contains none of the customary (albeit microscopically implausible) third order reactions, and all reverse reactions are allowed. The concentrations of X, Y, and Z are allowed to vary, and all others are assumed to be kept constant by compensating flows across the boundary. However, X,Y, and Z cannot pass the boundary. Let the concentrations be denoted by corresponding lower-case letters and forward and reverse reaction rate constants by k_i and k_{-i} (i=1,..., 5). The control parameter is the concentration of B, and for the other parameters we have chosen the values $k_1=k_3=1$, $k_{-1}=0.15$, $k_2=2.5$, $k_{-2}=10^{-4}$, $k_{-3}=0.275$, $k_4=1000$, $k_{-4}=10^{-3}$, $k_5=7000$, $k_{-5}=0.1$, $a=490$, $c=d=10$.

Again, the rate of entropy production due to the chemical reactions is given by (5.4). For the chemical potentials, we use the expression (5.6). The fact that affinities and reaction rates are zero in equilibrium yields the full affinity expressions. Each term in the sum in (5.4) is then clearly non-negative, and zero only if the individual reaction rate is zero.

The average rate of entropy production is calculated according to (5.10) and Ψ according to (5.11). The function $\Psi(b)$, both for this model and for a similar one, is shown in Fig. 5.1 (for full details, see Månsson [1985]). For b close to the lower Hopf bifurcation the average entropy production on a periodic trajectory is smaller than in the corresponding unique unstable stationary state, i.e., Ψ is negative, whereas the opposite occurs close to the upper Hopf bifurcation. Clearly, since this model exemplifies both positive and negative Ψ, there is no general rule that the

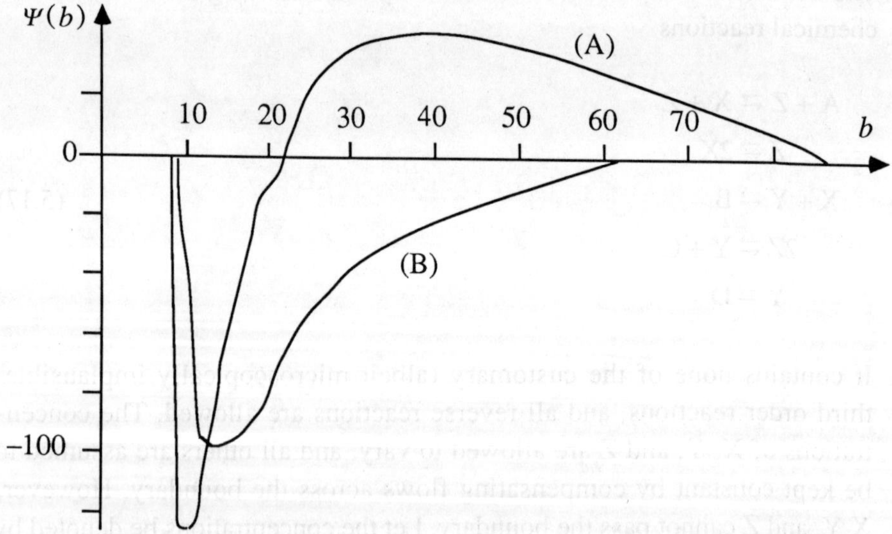

Fig. 5.1. The function $\Psi(b)$ for two different models; (A) is the model with three free concentrations described here, and (B) is another model (with four free concentrations) shown here to indicate that $\Psi(b)$ may have a definite sign in some cases.

entropy production is smaller in an oscillating mode than it would be if the system stayed in some of the available stationary states. In other words, a system may have a smaller entropy production in an oscillating mode for some combinations of driving forces, but not for all. A stationary state is not always favorable in terms of minimizing the entropy production, nor is an oscillating mode. Thus, it is proven by this counter-example that a general 'minimum entropy production principle,' based on time-averages over periodic oscillations, cannot be valid.

A model with period doublings and chaos

The natural next step is to investigate chemical oscillators with period-doubling bifurcations and chaos, in order to find out the behavior of the average rate of entropy production in periodic and chaotic modes as well as in coexisting stationary states [Lindgren and Månsson 1986].

In many cases reaction rates vary very little with small displacements in state space. Therefore, trajectories that are everywhere close in state space have roughly equal periods, modulo an integer factor. Close to a period doubling, this can be exploited to define an analogous function to Ψ, but comparing two averages of the kind (5.10) with one of them based upon an unstable periodic trajectory.

Willamowski and Rössler [1980] proposed a model which both satisfies the requirements of thermodynamics and exhibits chaotic behaviour. It has the reaction scheme

$$\begin{aligned} A + X &\rightleftarrows 2X \\ X + Y &\rightleftarrows 2Y \\ B + Z &\rightleftarrows 2Z \\ C + Y &\rightleftarrows D \\ X + Z &\rightleftarrows E \end{aligned} \tag{5.18}$$

All reverse reactions are allowed. As in the previous example, the concentrations of X, Y, and Z are allowed to vary, all others are assumed to be

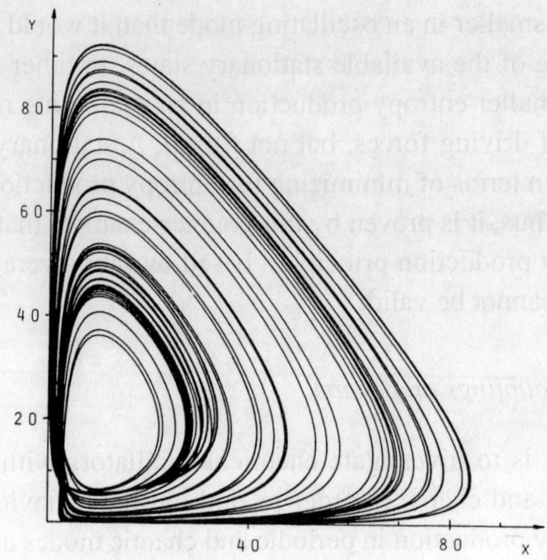

Fig. 5.2a. Projection onto the *x*-*y* plane of approximately fifty 'loops' of a chaotic trajectory for $a = 30.0$.

Fig. 5.2b. Poincaré map of a chaotic trajectory for $a = 30.0$ (intersections with the plane $x = 10$).

kept constant, etc. The control parameter is now the concentration of A, and the other parameters have the values $k_1=k_2=k_3=k_4=k_{-4}=k_5=k_{-5}=1$, $k_{-1}=0.25$, $k_{-2}=0.001$, $k_{-3}=0.5$, $b=16.5$, $c=10.0$, and $d=e=0.01$. Figure 5.2 shows a typical chaotic trajectory.

This system can have multiple stationary states for certain control parameter values. Figure 5.3 shows the Lyapunov exponents, (cf. Section 3.4.1), two of which are zero at each period doubling. There are several intervals where one exponent is positive, i.e., where chaotic modes exist. A comparison with the fixed-point analysis shows that there is also a locally *stable* stationary state for these parameter values. There are thus at least two basins of attraction in state space.

Again, the rate of entropy production due to the chemical reactions is given by (5.4), and the expression for the chemical potential is (5.6). The average rate of entropy production is calculated according to (5.10) for periodic trajectories. For chaotic trajectories a sampling procedure may be used, such that the value of (5.10) is calculated for non-overlapping time-

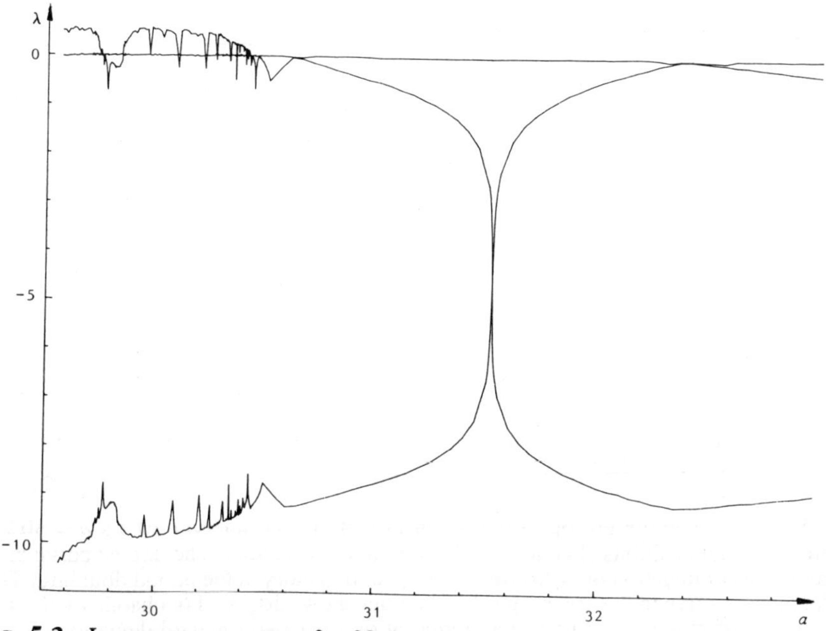

Fig. 5.3. Lyapunov exponents for $29.6 \leq a \leq 33.0$ showing several intervals with positive Lyapunov exponents, i.e., chaos.

intervals. Lindgren and Månsson [1986] used a time interval (τ in (5.10)) determined by following the trajectory (from the zeroth) to the 256th intersection with the *Poincaré plane*. This gave a distribution of σ_T values, and the mean of this distribution was regarded as the average rate of entropy production. The derivative $d\sigma_T/da$, while remaining positive, changes drastically at each period doubling. The model shows no discernible rule for determining whether the derivative increases or decreases at a period doubling. The sign of the quantity corresponding to $\Psi(\alpha)$ in (5.11) in the neighbourhood of a period doubling may be positive or negative. In fact, if the average entropy production on the unstable trajectory is approximated by an extrapolation (as indicated by the dashed curve in Fig. 5.4), the curve for $d\sigma_T/da$ shows that both cases occur in the present system.

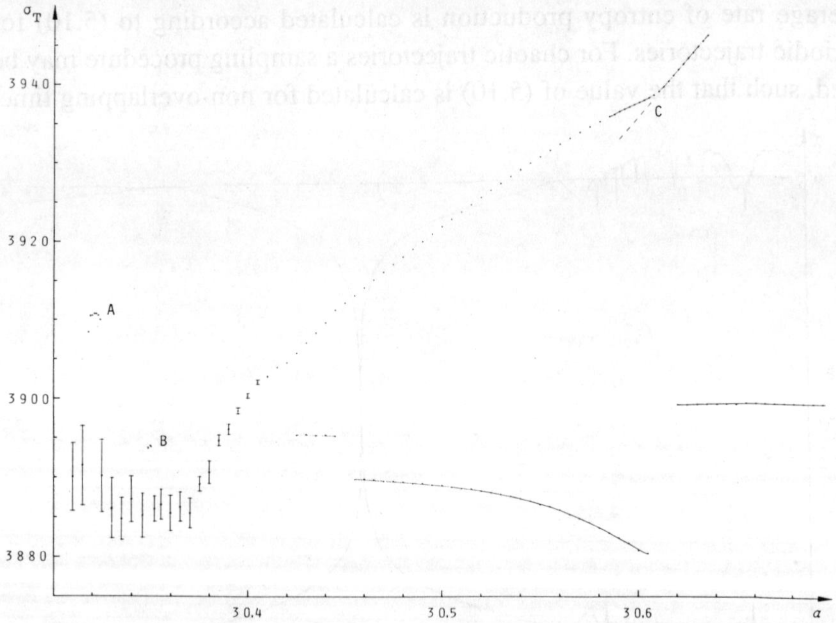

Fig. 5.4. The average entropy production rate σ_T in the interval $30.3 \leq a \leq 30.7$, where the system exhibits chaotic as well as periodic behaviour. The dashed curve at C represents a continuation of $\sigma_T(a)$ for the original trajectory at the period doubling. The discontinuous curve in the lower right-hand corner shows $d\sigma_T/da$. For chaotic modes the distribution of $\sigma_T(a)$ is shown as the midpoint of a bar two standard deviations long. Values of $\sigma_T(a)$ in two periodic windows are shown at A and B respectively.

5.1.3 *Temporal decomposition of contrast*

In Chapter 3 we introduced various ways of decomposing the contrast for a spatially distributed system. For a homogeneous chemical oscillator, a similar decomposition can be performed if the oscillations are periodic. Assume that the state of the system at time t is described by a concentration (molar fraction) vector $\boldsymbol{x}(t)$. For any two states $\boldsymbol{x}(t)$ and $\boldsymbol{x}'(t)$ (≥ 0) which fulfil the normalization condition ($\sum x'_i = \sum x_i = 1$) the contrast can be defined by

$$K[\boldsymbol{x}; \boldsymbol{x}'] = \sum_i x_i' \ln\left(\frac{x_i'}{x_i}\right). \tag{5.19}$$

As before, let the vector $\boldsymbol{x}^*$ be the stationary state and again assume that $\boldsymbol{x}(t) \in \mathrm{T}$, where T is a periodic trajectory with period τ. Then the *time-average* of the contrast over a period can be decomposed into two terms as follows:

$$\frac{1}{\tau}\int_0^\tau K[\boldsymbol{x}^*; \boldsymbol{x}'(t)] = \frac{1}{\tau}\int_0^\tau \sum_i x_i'(t)\left[\ln\left(\frac{x_i(t)}{\bar{x}_i}\right) + \ln\left(\frac{\bar{x}_i}{x_i^*}\right)\right] \mathrm{d}\,t =$$

$$= \frac{1}{\tau}\int_0^\tau K[\bar{\boldsymbol{x}}\,; \boldsymbol{x}'(t)] + K[\boldsymbol{x}^*; \bar{\boldsymbol{x}}'] = \overline{K[\bar{\boldsymbol{x}}; \boldsymbol{x}'(t)]} + K[\boldsymbol{x}^*; \bar{\boldsymbol{x}}'], \tag{5.20}$$

where bars denote time-averages. Thus, it consists of the contrast of the average state with respect to the stationary state, and the average of the contrast of the actual state with respect to the average state. The decomposition is analogous to the chemical-structural decomposition, but with temporal averaging instead of spatial.

5.2 Information Flows and Spatial Structures

Among the information-theoretical concepts introduced in Chapter 3 were information density and information flow in space *and* in scale (cf. (3.33)). These concepts were defined for a chemical system described by a set of normalized concentrations $c_i(\mathbf{x},t)$. We shall here describe a model for a chemical system which is spatially self-organizing. We have made dynamical simulations of this model in order to test the usefulness of the information-theoretical concepts. The conclusion is that the flow of contrast from large scale (chemical) through an intermediate scale (structure) into the microscopic scale (entropy production) is a picture, which quite well describes the structuring process.

5.2.1 *A model for a spatially self-organizing system*

For our chemical model [Gierer and Meinhardt 1972] we have chosen the following set of reactions:

$$\begin{aligned} P + X + A &\rightleftarrows P + 2X && \text{(a)} \\ 2X + B &\rightleftarrows 2X + Y && \text{(b)} \\ X &\rightleftarrows F && \text{(c)} \\ Y &\rightleftarrows G && \text{(d)} \\ P + Y &\rightleftarrows X && \text{(e)} \end{aligned} \tag{5.21}$$

It conserves the quantity

$$N = N_B + N_X + N_Y + N_F + N_G \,. \tag{5.22}$$

The reaction (a) is self-catalytic, and both reactions (a) and (b) are catalysed by X, which is called the 'activator' of the system. Reaction (e) is assumed to be fast, so that equilibrium is immediate. Then the concentration c_P of P is proportional to c_X but inversely proportional to c_Y. Since P is a catalyst in reaction (a) where the activator X is produced, Y is an 'inhibitor' of the process. Both X and Y are produced from B, which is assumed to be exergy rich and therefore serves as a 'fuel' in the process. Similarly F and G are 'waste products' from the decay of X and Y in reactions (c) and (d). All components are allowed to diffuse in the system, and it is essential that the diffusion constant for the activator is smaller than that of the inhibitor.

The system is assumed to be one-dimensional with periodic boundary conditions. Initially all components are distributed homogenously in the system, but the homogeneity is unstable against fluctuations. In the simulation we let the concentration of the activator fluctuate in space and time. In the beginning the system is sensitive to fluctuations which initiate the development of a spatial structure.

The processes (5.21) have been studied in two different cases [Eriksson and Lindgren 1987]. In the first case the system is assumed to be closed, i.e., no inflow or outflow of chemical components is allowed. The initial state is prepared with high concentration of the fuel and low concentrations of the waste products. In the second case the system is open, and the inflow of B and the outflows of F and G are such that their concentrations c_B, c_F and c_G are held constant in space and time. The formation of a typical spatial pattern for the closed case is shown in Fig. 5.5.

In both cases the structural contrast k_{str} and the entropy production rate σ were calculated. The major part of σ is due to chemical reactions. Thus σ indicates the rate of consumption of chemical contrast. The other part, due to diffusion within the system, is much smaller.

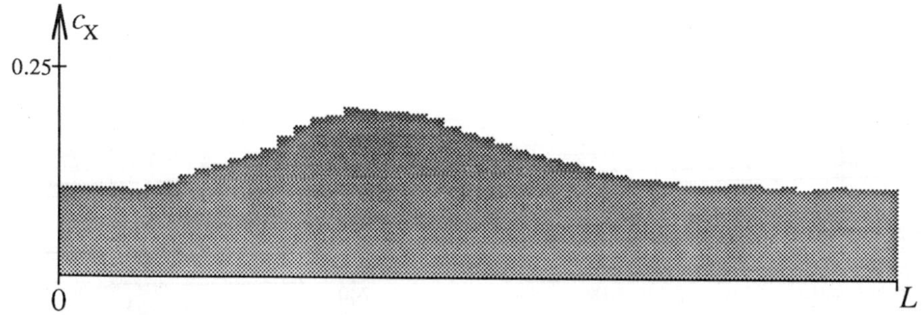

Fig. 5.5. The concentration of the activator as a function of spatial coordinate, at a time when a spatial structure has emerged. The system, of length L, is divided into 50 cells.

In the case of a closed system with no external supply of B (Fig. 5.6a), there is first a construction phase, during which k_{str} increases, followed by a decay phase during which k_{str} and k_{chem} both tend to zero. During the con-

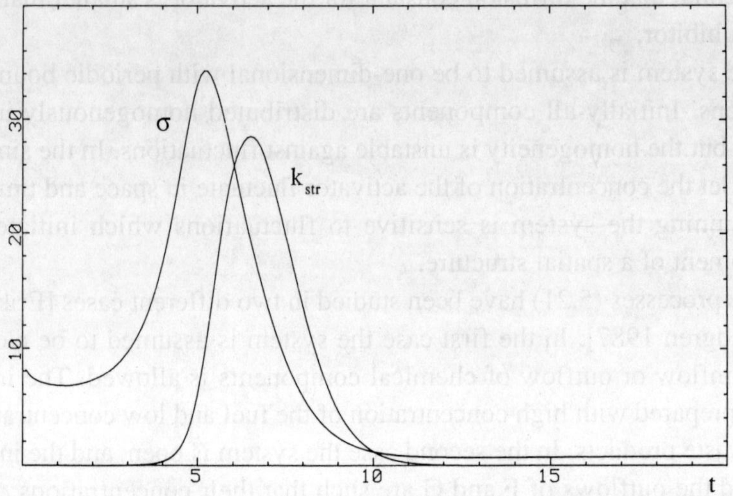

Fig. 5.6a. The build-up of structure leads to a maximum in entropy production. Since the system is closed, the structure must decay.

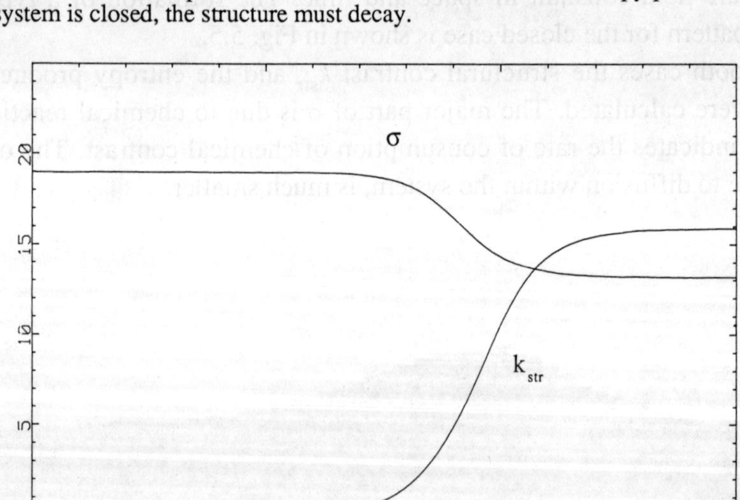

Fig. 5.6b. The open system reaches a stationary level in k_{str}, which actually reduces the consumption of exergy.

struction phase chemical exergy is used to build up the structure, leading to a maximum in the entropy production.

In the second case, the open system (Fig. 5.6b), there is a construction phase followed by a stationary maintenance phase. The spatial pattern in the maintenance phase is similar to the most structured pattern in the closed case. The structure built up actually reduces the consumption of chemical exergy. (The initial state was chosen so that it would be stationary in the absence of spatially inhomogeneous fluctuations.)

5.2.2 *Flows of information during structure formation*

We shall now make a detailed analysis of the model of the closed chemical self-organizing system described above, based on information-theoretical concepts from Chapter 3.

We study the decomposition of the total contrast $K(t)$ into contributions $k(r,x,t)$ from spatial coordinate x and resolution r, cf. (3.25-26). In Chapter 3 it was assumed that the system was embedded in an equilibrium environment so that for $r \to \infty$, $k(r,x,t) \to 0$. Here the system is closed so that the contribution to $k(r,x,t)$ at infinite r is equal to the chemical contrast $k_{\text{chem}}(t)$. During the process of structure formation (and decay), the information flow $j_r(r,x,t)$ (3.29) between different levels of resolution r, and the information flow $j_x(r,x,t)$ (3.32) in space, can be calculated. The local increase of structural contrast $\partial k(r,x,t)/\partial t$ is related to j_r and j_x through the continuity equation (3.30)

$$\dot{k}(r,x,t) = r\frac{\partial}{\partial r}\, j_r(r,x,t) - \frac{\partial}{\partial x}\, j_x(r,x,t)\,. \tag{5.23}$$

For $r \to \infty$, there is a positive flow $j_r(r,x,t)$ in the direction of smaller r, due to the chemical contrast at infinite r,

$$j_r(r,x,t) = -\,\dot{k}_{\text{chem}}(t), \quad \text{all } x, t\,; \quad r \to \infty\,. \tag{5.24}$$

If there were no spatial dependence in the system, this flow would be con-

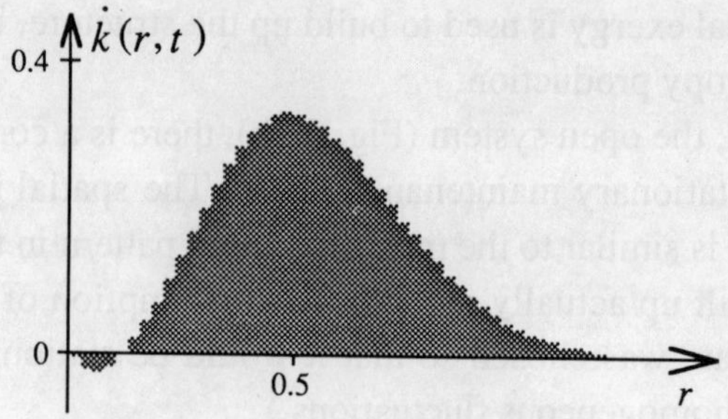

Fig. 5.7. The growth of contrasts as a function of resolution shows a build-up of structure for r=0.5, while there are smaller structures that decay. The time t=t_1, is the same as in in Fig. 5.5.

stant in x, and it would go from the macroscopic scale at $r = \infty$, and enter the microscopic scale as entropy production (3.27-28), $j_r(0,x,t) = \sigma(t)$.

In the closed spatially self-organizing system described in the previous section, we start with a homogenous distribution of concentrations. As we saw, this system can amplify fluctuations, and develop a spatial structure. The noise gives rise to a small information flow from the microscopic to the macroscopic level, which leads to small changes in the flows $j_r(r,x,t)$ and $j_x(r,x,t)$ for small r. These changes can be amplified and moved to higher scales r, where the spatial structure appears.

By integrating $\dot{k}(r,x,t)$ over x, we get the contrast growth $\dot{k}(r,t)$, which shows how structures of different size r develop. In Fig. 5.7, $\dot{k}(r,t)$ is plotted as a function of r, for a time $t_1 = 3$ in the beginning of the structure formation. There is a maximum showing the build-up of structure, but there is also a small minimum for $r = 0.06$ with a negative $\dot{k}(r,t)$, which shows that there are structures of small size which decay.

In Fig. 5.5 the concentration of the activator $c_X(x,t)$ was shown for the time t_1. For the level of perfect resolution $r = 0$, j_r shows the spatial dependence of the entropy production rate (Fig. 5.8), which is high in theconcentration peak since X is a catalyst of the fuel consuming reactions (5.21a-b). In comparison, the rate of entropy production due to the diffusion is so small that it is not visible. Each term in the continuity equation is zero for $r = 0$.

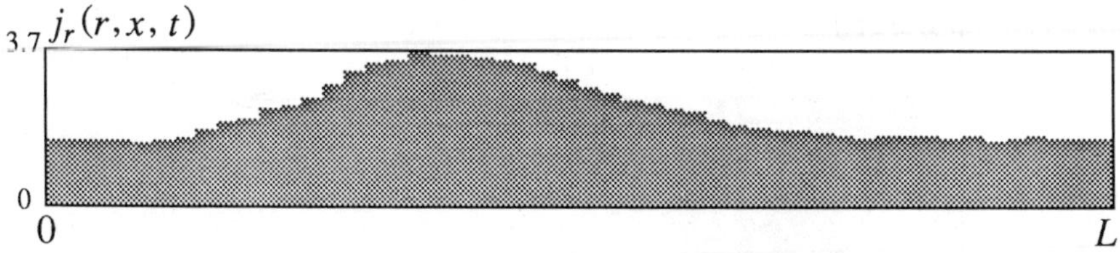

Fig. 5.8. The flow j_r, for r=0 and t=t_1, shows the local entropy production in the system.

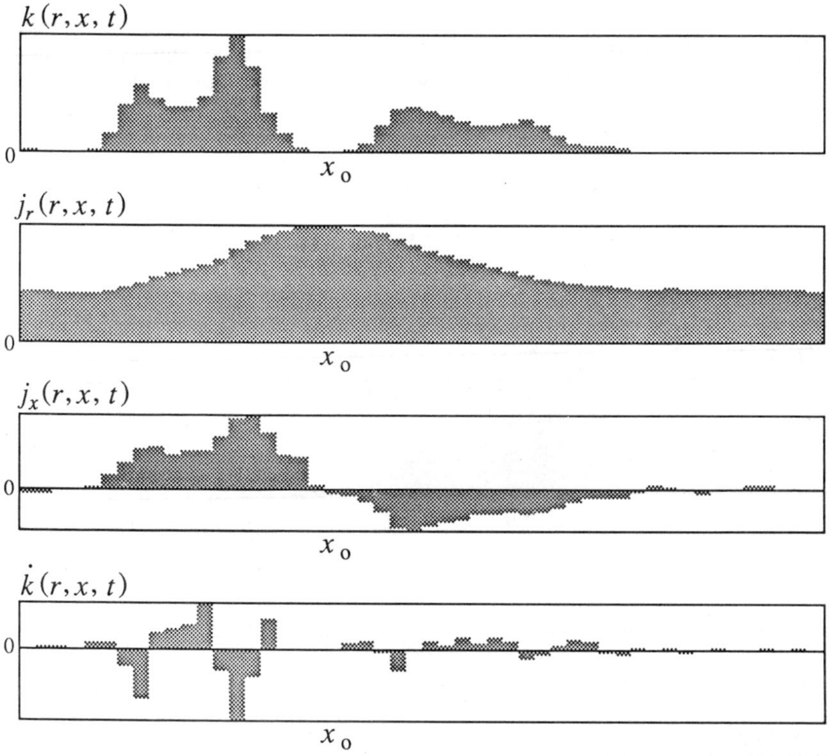

Fig. 5.9. The spatial dependence of contrasts and flows, for r=0.06 and t=t_1, corresponding to the minimum in Fig. 5.7. The contrast k and its time derivative clearly shows that there are structures, overlaid on the main concentration peak, which are decaying.

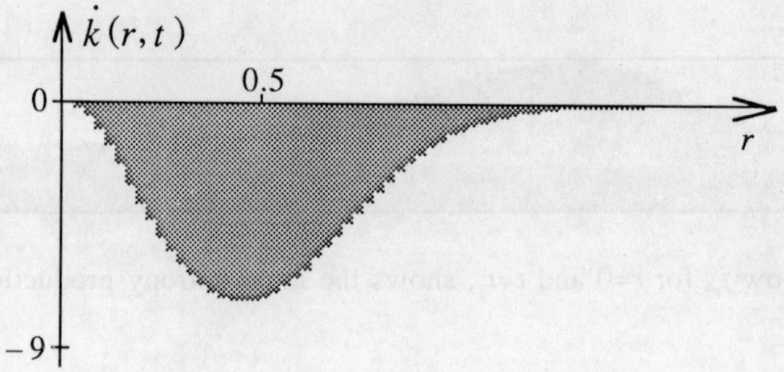

Fig. 5.10. The time derivative of the contrast as a function of resolution for a time t_2. It shows that the system is in a decay phase of the time-development depicted in Fig. 5.6a.

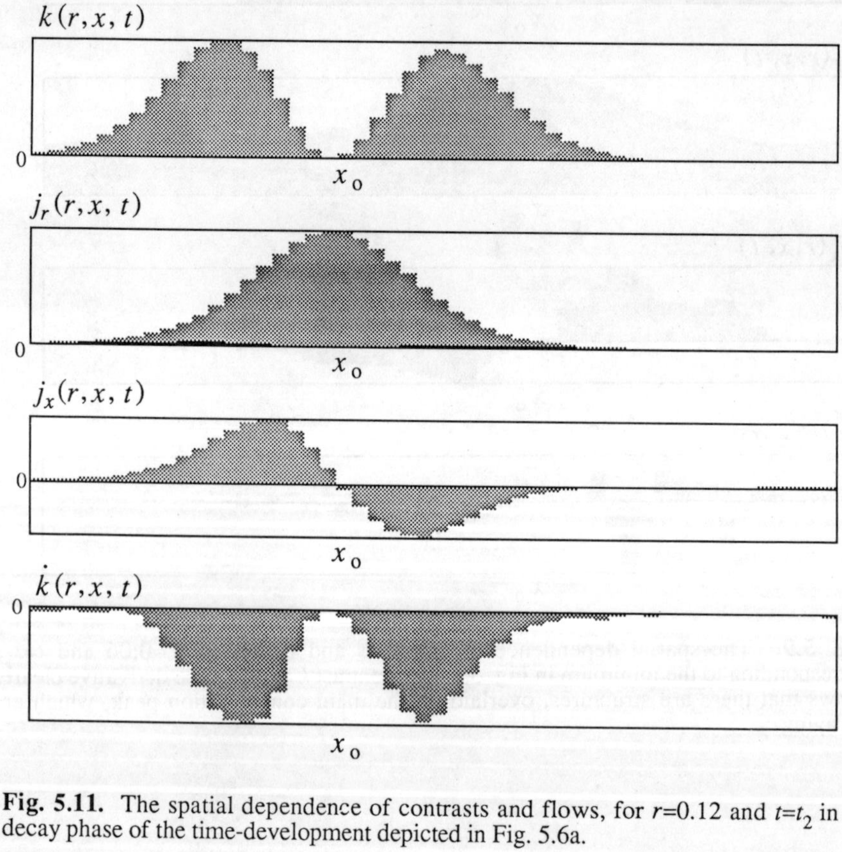

Fig. 5.11. The spatial dependence of contrasts and flows, for r=0.12 and t=t_2 in the decay phase of the time-development depicted in Fig. 5.6a.

In Fig. 5.9, the local contrast $k(r,x,t)$, the flows $j_r(r,x,t)$ and $j_x(r,x,t)$, and the time derivative of $k(r,x,t)$ are plotted as functions of x, for $t=t_1$ and $r = 0.06$ (corresponding to the minimum in Fig. 5.7). The contrast $k(r,x,t)$ and its time derivative clearly show the presence of decaying structures that are smaller than the major concentration peak. At the position x_0 of the concentration maximum, we have that $-\partial j_x/\partial x > 0$ and $\dot{k}=0$, which, according to the continuity equation, give $r\ \partial j_r/\partial r < 0$. In the surroundings of x_0 (at the slopes of the concentration peak), the contrast flow from higher r is partly transformed into $j_x(r,x,t)$ flowing towards the center, x_0.

Since the system is closed, the structures that have been built up must also decay. In Fig. 5.10, $\dot{k}(r,t)$ is shown for a time t_2 in the decay phase of the process. There is a negative minimum corresponding to the maximum of Fig. 5.7, which shows how the structure decays. The spatial dependence of the different flows is shown in Fig. 5.11.

5.3 Biochemical Systems, Information, and Hypercycles

Above, we investigated a few thermodynamical evolution criteria for some kinds of chemical self-organizing systems. We shall now turn to the most interesting and difficult subject, namely the living systems. The set of living systems constitute only a subset of the set of all self-organizing chemical systems, but it is the most important one. We shall nevertheless limit ourselves to only review some pertinent thermodynamical and information-theoretical aspects of the evolution of living systems. (Our discussion relies heavily on Küppers' exposition [Küppers1983].) We shall address some aspects of the question "What is life?" and outline parts of a physico-chemical theory of the origin of life. (See also the comprehensive work of Miller [1978] and, e.g., Peacocke [1983].) For a formulation and analysis of such problems in physical terms, information theory, including statistical mechanics, turns out to be the most relevant available theory.

We may here note that the entropy decrease associated with building a living organism from its chemical building blocks is very small (see, e.g.,

Blumenfeld [1981]). According to such traditional thermodynamical concepts as the entropy, the amount of 'ordering' of any biological system is not significantly different from that of a stone of the same weight. Also, the difference in structural information between living and non-living systems may be very small.

It is customary to take the following three properties as characteristic of all living organisms:

1. Metabolism (from the greek *μεταβαλλειν*, meaning change or exchange). This means that they are open systems, i.e., they exchange matter and energy with other systems.
2. Self-reproduction. They are not likely to reappear in a random production process.
3. Mutability. Without mutability, there can be no evolution.

As we saw in Chapter 4, and will see below, these properties are suitable for information-theoretical analysis, although the existing concepts are not able to cover all essential aspects.

We can discern (at least) the following three stages in the emergence of living matter (from Küppers [1983]):

I. Chemical evolution; the non-instructed synthesis of biological macromolecules. In this stage all the chemicals are formed which are needed for the nucleation of life, in particular the nucleic acids and the proteins.

II. Molecular self-organization, i.e., *instructed* synthesis of biological macromolecules and organization into self-reproducing genetic systems. This is the stage in which the actual transition from non-living to living matter occurs.

III. Biological evolution, including the whole development from primitive genetic systems to complex multicellular organisms.

In stage I the important process can be described as the accumulation of tied-up exergy in molecules. In stage II the most important process is the

action, here denoted ΔG_{ij} for pairing of bases i and j. For incorporation of, e.g., the nucleotide A (in RNA) we have under certain simplifying assumptions (see Küppers [1983], Sect. 7.3, for details),

$$q_A = \left(\frac{s_{AU}^2}{\left(\sum_{i=A,U,G,C} s_{Ai} \right) \left(\sum_{i=A,U,G,C} s_{Ui} \right)} \right)^{\frac{1}{2}}, \qquad (5.33)$$

where we have introduced the pair stability constants

$$s_{ij} = \exp\left(-\frac{\Delta G_{ij}}{RT} \right). \qquad (5.34)$$

Obviously, for finite ΔG_{ij} and non-zero temperature T, the q_α's (and consequently Q_I) are less than unity.

For the complementary base pairs AU and GC, one may use $s_{AU}=s_{GU}=$ $=10$ and $s_{GC}=100$, all other $s_{ij}=1$, leading to $q_A=q_U=0.59$ and $q_G=q_C=0.93$. This is consistent with the generally accepted view that G and C reproduce with an appreciably higher fidelity than A and U. For typical Θ values in the range 2-200, the threshold relation (5.32) yields longest chain lengths of 14-106 bases. (Details of the reasoning behind these estimates were given by Eigen and Schuster [1979].) This means that without the help of complex molecular copying-machines not much more than 100 nucleotides in a nucleic acid can be correctly copied under selection pressure.

Clearly, an expansion of information capacity to the typical levels present in organisms demands a significant improvement in copying accuracy. Living systems contain molecular 'copying machines' (similar to the Brownian computer discussed in Section 4.5.1), the polymerase enzymes (enzyme = protein with the ability to catalyze chemical reactions), which are able to decrease the error rate considerably. On the other hand, the genetic code for such polymerases requires much more than 100 symbols. It is therefore clear that some other kind of information storage and accumulation mechanism was necessary to get over the information barrier, or information crisis gap, associated with the intermediate range of chain lengths of stage II. Higher-order functions, such as 'proof-reading,' which can give a further boost in accuracy (to $1-q=10^{-6}$-10^{-9}), require even more symbols (since the 'blue-print' for the 'machinery' performing such a function must also be stored) but also allow much more information storage. Historically, they occured in stage III.

5.3.2 *Accumulation of information: the hypercycle*

The crucial assumption in the derivation above turns out to be that of *kinetical independence* of the information carriers. One way of abandoning this assumption leads to a mechanism, the hypercycle, which, according to Eigen and Schuster [1979], constitutes one of the missing links in the prebiotic evolution from simple self-replicating elements with enzyme-free copying mechanism (stage I) to the highly organized early RNA (stage III).

Assuming constant total population as before we can write

$$\dot{x}_i = \Gamma_i(x;\Lambda) - \frac{x_i}{C}\sum_j \Gamma_j(x;\Lambda), \tag{5.35}$$

where Γ_i is the growth function and Λ is a set of parameters. Writing the growth function on the form

$$\Gamma_i = k_i + \sum_j k_{ij}\, x_j + \sum_{j,l} k_{ijl}\, x_j\, x_l + \ldots, \tag{5.36}$$

we can define its degree d_i as the highest power appearing in (5.36). If all d_i-values are identical we call the system *homogeneous*, otherwise it is inhomogeneous. For independently growing species we have

$$\Gamma_i = k_i\, x_i\,, \tag{5.37}$$

i.e., a homogeneous system very similar to (5.27). An acyclic catalytic chain has

$$\begin{cases} \Gamma_1 = k_1 x_1 ; \\ \Gamma_i = k_i x_i + k_i' \, x_i \, x_{i-1} , \end{cases} \tag{5.38}$$

i.e., it is an inhomogeneous system. Fixed-point analysis of this system indicates that it is not likely to result in stabilisation of information.

A *hypercycle* is a catalytically-coupled chain of N self-replicating species (Fig. 5.12), which are coupled together by a *closed* loop of reactions. It is perhaps the simplest chemical reaction network model which can exhibit the characteristic properties of living organisms listed above.

Generally, the self-reproduction of each species can be catalyzed by any other species in the loop. A simple mathematical model for such a system is given by

$$\Gamma_i = k_i \prod_{j=1}^{N} x_j^{d_{ij}} , \tag{5.39}$$

where $d_{ij} = 0$ or 1 and where the N variables are taken in cyclic order. We

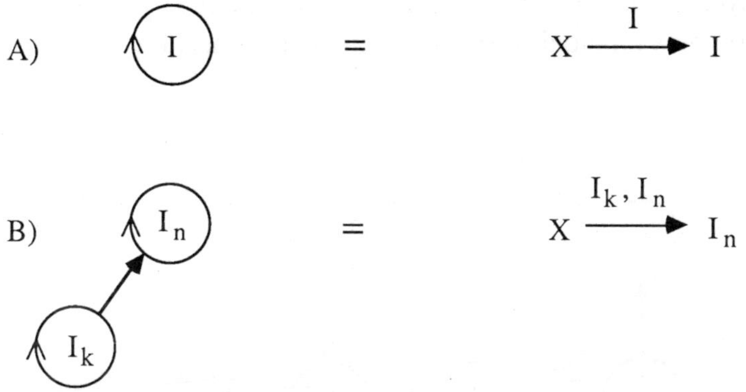

Fig. 5.12. Definition of the symbols for: A) a self-replicative unit; B) two coupled self-replicative units. X denotes the reactants.

assume that it is homogeneous of degree d, where

$$d = \sum_j d_{ij} \,. \tag{5.40}$$

For a hypercycle,

$$N \geq d \geq 2, \tag{5.41}$$

must be fulfilled. A particular hypercycle can be described by the matrix $D=\{d_{ij}\}$, e.g., the elementary hypercycle ($d = 2$) with N=6 is described by

$$D = \begin{pmatrix} 1 & 1 & 0 & 0 & 0 & 0 \\ 0 & 1 & 1 & 0 & 0 & 0 \\ 0 & 0 & 1 & 1 & 0 & 0 \\ 0 & 0 & 0 & 1 & 1 & 0 \\ 0 & 0 & 0 & 0 & 1 & 1 \\ 1 & 0 & 0 & 0 & 0 & 1 \end{pmatrix} . \tag{5.42}$$

This hypercycle is also shown in Fig. 5.13. Clearly, large hypercycles may have very complex topologies. We may note that connection of two hyper-

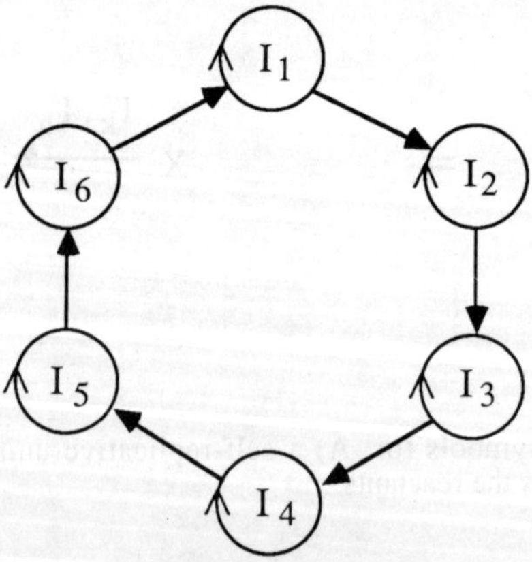

Fig. 5.13. The elementary hypercycle, described by (5.42).

cycles results in one large hypercycle.

The following elementary hypercycle is said to resemble the protein biosynthetic cycle in organisms [Eigen and Schuster 1979, Küppers 1983], and is therefore regarded as a possible precursor of the living cell: N is an even number; the species are $N/2$ polynucleotides I_i (where i is odd, say) and $N/2$ polypeptides E_j (j even). Each polynucleotide consists of an autonomously self-reproducing pair of nucleic acids with complementary base-pairing. The reproduction of each I_i is catalyzed by E_{i-1}, and each E_j is the translation product of I_{j-1}, and, to close the circle, E_N acts upon I_1.

It is often convenient to scale the dynamical system (5.35) so that it is defined on the concentration N-simplex

$$X^{(N-1)} = \left\{ x_i \in \mathrm{R}^N : x_i \geq 0, \sum_{j=1}^{n} x_j = 1 \right\}. \tag{5.43}$$

Fixed-point analysis shows that the system (5.39) has a single interior stationary point $\boldsymbol{x}^*=(1,\ldots,1)/N$. There are N fixed points lying at the corners of $X^{(N-1)}$. There may be further sets of fixed points on the periphery of $X^{(N-1)}$. Linear stability analysis shows that the point $\boldsymbol{x}^*$ is exponentially stable for $N = 2$ and 3, linearly stable for $N = 4$, and unstable for $N \geq 5$. For $N = 2$, 3, and 4, a Lyapunov function argument shows that $\boldsymbol{x}^*$ is globally attracting in the interior of $X^{(N-1)}$. For $N = 5$ there is a stable interior closed trajectory. There are indications from numerical simulations that for $N \geq 5$ there is at least one stable periodic solution with a large basin of attraction. It seems to be fairly well established that all elementary hypercycles show the same dynamic behaviour; each possesses exactly one attractor in the interior of $X^{(N-1)}$. This demonstrates their ability to maintain co-existence of *all* of their species.

The advantage of the hypercycle is that the individual information carriers promote each other's growth. Instead of having to fight with mutants, as the master sequence of a quasi-species has to do, a species *collaborates* with other species in a way which ensures its own survival. The information decay process which led to the condition (5.32) is thereby circumvented, and the development of longer polynucleotide chains becomes possible.

An obvious generalization of the analysis above is to consider networks of interacting hypercycles and to study selection and evolution processes in such systems.

Let us consider a hypercycle of the kind desribed by (5.39) and 5.42). It may result from a set of reactions, adding up to

$$A_i + X_i + X_{i-1} \rightarrow B_i + 2X_i + X_{i-1} \ (+Q_i), \tag{5.44}$$

where Q_i is the dissipated heat. Note that, in the following, the variables J_i, L_i, and D_i denote import, export, and dissipation flow respectively, cf. (4.83). We assume now that the back reaction is relatively unimportant and can be neglected. We further assume that the concentrations are constant over the volume of interest. In terms of the terminology of Section 4.6, we may think of A_i as imported raw material for X_i and B_i as a rest product exported from the reaction. If the concentration of A_i is a_i, then

$$\begin{cases} J_i = L_i = k_i a_i x_i \, x_{i-1}; \\ D_i = Q_i J_i \, . \end{cases} \tag{5.45}$$

The internal turnover of X_i is

$$M_{ii} = k_i a_i \, x_i \, x_{i-1} + k_{i+1} a_{i+1} x_{i+1} x_i \, , \tag{5.46}$$

and

$$M_{ij} = 0, \ \ i \neq j. \tag{5.47}$$

Assume now that we have competing hypercycles (labelled (j)), using the same inputs A_i. Then

$$\begin{cases} J_i^{(j)} = L_i^{(j)} = k_i^{(j)} a_i x_i^{(j)} x_{i-1}^{(j)}; \\ D_i^{(j)} = Q_i^{(j)} J_i^{(j)}; \\ M_{ii}^{(j)} = k_i^{(j)} a_i x_i^{(j)} x_{i-1}^{(j)} + k_{i+1}^{(j)} a_{i+1} x_{i+1}^{(j)} x_i^{(j)}. \end{cases} \tag{5.48}$$

Very quickly the most efficient hypercycle will win over all the others and monopolize the inputs A_i. The concentrations of A_i will stabilize on a level where there is balance with the inflow from the surroundings. The total amount of material flows increases up to this maximum. At the same time there is a strong specialization to the specific channels determined by the winning hypercycle. Using (4.97) to (4.99), we see that the development capacity C and the ascendency A grow through the competition, but the redundancy R remains relatively small. If we disaggregate the system by dividing it into volume cells and consider diffusion between the cells, then it is clear that also R may increase through the competition as a result of the increase in the total turnover. The resulting redundancy corresponds to the possibility of the system to continue the specialization geometrically, i.e., to form spatial patterns.

We conclude that the evolutionary measures of Section 4.6, which were introduced in ecology, have a wider application, and in particular they seem to suit the competition between hypercycles quite well.

6. QUANTUM SYSTEMS

Quantum mechanics is fundamental to modern physics, in particular to the physics of material structure. It is often stated that quantum mechanics is in itself non-deterministic. This is not true: equations of motion in quantum mechanics are as deterministic as their classical counterparts. What is true is that the prediction of the outcome of a measurement in general can be done only in terms of probabilities. This has led to the formalism of quantum statistics, in terms of which also information theory can be formulated. As in the classical case a stochastic influence on a quantum system leads to irreversible dynamics.

In this chapter we shall review the formalism of quantum statistics, and then apply it, first to obtain the dynamics of a system in interaction with a stochastic field, and then to extend some geometrical aspects of information theory, developed in Chapter 3 to the phase-space of a set of oscillators. Then the quantum counterpart of computation thermodynamics, discussed in Chapter 4, is reviewed. The chapter ends with a general model of open quantum-mechanical systems, which in the thermodynamic limit leads to a bifurcating behaviour, similar to that of self-organizing systems. The model therefore provides a theory for quantum measurement, based entirely on quantum mechanics and statistics.

This chapter is not very closely connected to the evolutionary perspective of the book, described in the Introduction. However, it is an integral part of the program to tie information theory and statistical physics closer together.

6.1 Quantum Statistics

In statistical quantum mechanics, a macrostate is described by a density operator ρ usually considered as a matrix operator, see, e.g., [Messiah 1961, Chapter 8:IV]. If $|0\rangle, |1\rangle, |2\rangle, \ldots$ form an orthonormal basis in the Hilbert space of microstates, $\langle k|\rho|k\rangle$ is interpreted as the probability for the microstate $|k\rangle$ in the macrostate ρ, and we require

$$\begin{cases} <k|\rho|k> \geq 0, \quad \text{all } k; & (6.1) \\ \operatorname{Tr}\rho = \sum_k <k|\rho|k> = 1. & (6.2) \end{cases}$$

In the special case when ρ describes a system in a pure quantum state, say $|\psi>$, we have

$$\rho = |\psi><\psi|. \tag{6.3}$$

and the probability to observe the system in the state $|k>$ is

$$<k|\rho|k> = |<k|\psi>|^2, \tag{6.4}$$

as in ordinary quantum mechanics.

Because of (6.1), ρ is a positive semidefinite operator, and it may be diagonalized by a unitary operator U ($UU^\dagger = 1$). In the basis $U^\dagger|k>$,

$$<k|U\rho U^\dagger|l> = p_k\,\delta_{kl}; \quad \sum_k p_k = 1, \tag{6.5}$$

where p_k ($k = 1, 2,\ldots$) represent ordinary probabilities of non-interfering states. The mean value of any arbitrary operator A in the macrostate ρ is

$$<A> = \sum_k p_k <k|UAU^\dagger|k> =$$
$$= \sum_{k,l} <k|U\rho U^\dagger|l><l|UAU^\dagger|k>, \tag{6.6}$$

i.e.,

$$<A> = \operatorname{Tr}[\rho A]. \tag{6.7}$$

Similarly the definition of entropy can be generalized from the representation (6.5),

$$S[\rho] = -\operatorname{Tr}[\rho \ln \rho]. \tag{6.8}$$

The Rényi entropy (see Section 1.4) can be defined as

$$S_\alpha[\rho] = \frac{1}{\alpha - 1} \ln \operatorname{Tr}[\rho^\alpha]. \tag{6.9}$$

For a standard macroscopic state ρ_o the Kullback information or contrast can be defined as

$$K[\rho_o; \rho] = \operatorname{Tr}[\rho\,(\ln \rho - \ln \rho_o)]. \tag{6.10}$$

It may be generalized to the Rényi case,

$$K_\alpha[\rho_o; \rho] = \frac{1}{\alpha - 1} \operatorname{Tr}[\rho\, e^{(\alpha-1)(\ln \rho - \ln \rho_o)}]. \tag{6.11}$$

If $\boldsymbol{f} = (f_1, \ldots, f_r)$ is a set of operators with prescribed expectation values, i.e.,

$$<\boldsymbol{f}> = \operatorname{Tr}[\rho \boldsymbol{f}] = \boldsymbol{F}\,, \tag{6.12}$$

then the macrostate ρ which maximizes the entropy (6.8) under the constraints (6.2) and (6.12) is the quantum Gibbs macrostate,

$$\rho = e^{-\mu(\boldsymbol{y}) - \boldsymbol{y}\cdot\boldsymbol{f}}, \tag{6.13}$$

where $\boldsymbol{y} = (y_1, \ldots, y_r)$ are Lagrangian multipliers and $\mu(\boldsymbol{y})$ is the logarithm of the partition function,

$$\mu(\boldsymbol{y}) = \ln \operatorname{Tr}[e^{-\boldsymbol{y}\cdot\boldsymbol{f}}]. \tag{6.14}$$

The condition for determining $\boldsymbol{y}$,

$$\boldsymbol{F} = -\frac{\partial \mu(\boldsymbol{y})}{\partial \boldsymbol{y}}, \tag{6.15}$$

is equivalent to (6.12).

For a system consisting of N similar, but distinguishable, subsystems (e.g., occupying different space regions), assuming $\boldsymbol{f}$ to be additive,

$$\boldsymbol{f} = \sum_k \boldsymbol{f}_k, \tag{6.16}$$

we have

$$\begin{cases} \rho = \mathrm{e}^{-N\mu(\boldsymbol{y}) - \boldsymbol{y}\cdot\boldsymbol{f}}; \\ \mu(\boldsymbol{y}) = \ln \mathrm{Tr}_k\,[\mathrm{e}^{-\boldsymbol{y}\cdot\boldsymbol{f}_k}], \\ \dfrac{\partial \mu(\boldsymbol{y})}{\partial \boldsymbol{y}} = -\dfrac{1}{N}\boldsymbol{F}. \end{cases} \tag{6.17}$$

6.1.1 *Coherent state formalism*

A useful model system to study is a set of n harmonic oscillators, described by the operators

$$\begin{cases} \boldsymbol{a} = (a_1, \ldots, a_n); \\ [a_\mu, a_\nu] = [a_\mu^\dagger, a_\nu^\dagger] = 0; \\ [a_\mu, a_\nu^\dagger] = \delta_{\mu\nu}. \end{cases} \tag{6.18}$$

The n operators

$$N_\mu = a_\mu^\dagger a_\mu \quad \text{(no summation)}, \tag{6.19}$$

then form a maximal set of commuting Hermitean operators. The eigenvectors are so-called *Fock states*

$$| m_1, m_2,\ldots, m_n > = \prod_{\mu=1}^{n} (m_\mu!)^{-1/2} (a_\mu^\dagger)^m | 0 >, \tag{6.20}$$

where $| 0 >$ is the ground state vector which satisfies

$$\boldsymbol{a} | 0 > = 0. \tag{6.21}$$

Then

$$\begin{cases} N_\mu | m_1,\ldots, m_\mu,\ldots, m_n > = m_\mu | m_1,\ldots, m_\mu,\ldots, m_n >; \\ < m_1',\ldots, m_n' | m_1,\ldots, m_n > = \prod_{\mu=1}^{n} \delta_{m_\mu' m_\mu}. \end{cases} \tag{6.22}$$

The class of states in quantum mechanics which come closest to the classical states are the coherent states [Klauder and Skagerstam 1985]. For any complex n-dimensional vector $\boldsymbol{z}$ we can define a unitary operator (star denotes complex conjugation),

$$U(z) = e^{\boldsymbol{a}^\dagger \cdot \boldsymbol{z} - \boldsymbol{z}^* \cdot \boldsymbol{a}} = e^{-\frac{1}{2} \boldsymbol{z}^* \cdot \boldsymbol{z}} e^{\boldsymbol{a}^\dagger \cdot \boldsymbol{z}} e^{-\boldsymbol{z}^* \cdot \boldsymbol{a}}. \tag{6.23}$$

From this we may construct the coherent state

$$| \boldsymbol{z} > = U(z) | 0 >, \tag{6.24}$$

which satisfies

$$\begin{cases} \boldsymbol{a} | \boldsymbol{z} > = \boldsymbol{z} \; | \boldsymbol{z} >; \\ < \boldsymbol{z} | \boldsymbol{a}^\dagger = \boldsymbol{z}^* < \boldsymbol{z} |; \\ < z_1 | z_2 > = e^{-\frac{1}{2}(z_1^* - z_2^*)\cdot(z_1 - z_2)} e^{\frac{1}{2}(z_1^* \cdot z_2 - z_2^* \cdot z_1)}. \end{cases} \tag{6.25}$$

The pure state density operator $|z> <z|$ then corresponds to a delta-function probability distribution in the classical case. However, now there is a spread around the mean value

$$\mathrm{Tr}\Big(|z><z|\,\frac{1}{2}\,[(a_\mu^\dagger - z_\mu^*)\,(a_\nu - z_\nu) + (a_\nu - z_\nu)(a_\mu^\dagger - z_\mu^*)]\Big) = \frac{1}{2}\,\delta_{\mu\nu}\,. \quad (6.26)$$

This is the minimum uncertainty in phase space allowed by the Heisenberg relation.

For a single mode, as we saw in the photon case (cf. (2.107)), the state of total incoherence for given expectation value B of the number operator $a^\dagger a$, is described by the density operator

$$\rho(B) = \sum_{n=0}^{\infty} \frac{1}{B+1}\left(1+\frac{1}{B}\right)^{-n} |n><n| = \frac{1}{B+1}\,\mathrm{e}^{-a^\dagger \ln(1+B^{-1})a} \quad (6.27)$$

which maximizes the entropy (6.8) under the constraints

$$\begin{cases} \mathrm{Tr}\,\rho = 1; \\ \mathrm{Tr}(\rho a^\dagger a) = B\,. \end{cases} \quad (6.28)$$

In terms of coherent states this may be seen as a Gaussian distribution centered around the ground state,

$$\rho(B) = \frac{1}{B\,\pi}\int \mathrm{d}^2 z\,\mathrm{e}^{-|z|^2/B}\,|z><z|. \quad (6.29)$$

The value of the entropy is

$$S(B) = (B+1)\ln(B+1) - B\ln B\,. \quad (6.30)$$

A more general set of density operators is the following [Glauber 1963]

$$\begin{cases} \rho(p) = \dfrac{1}{\pi}\displaystyle\int \mathrm{d}^2 y\,p(y)\,|y><y|; \\ p(y) \geq 0; \quad \dfrac{1}{\pi}\displaystyle\int \mathrm{d}^2 y\,p(y) = 1. \end{cases} \quad (6.31)$$

In particular if $p(y)$ is a Gaussian

$$p(y) = \frac{1}{B}\,\mathrm{e}^{-|y-z|^2/B}; \quad B > 0, \quad (6.32)$$

we have a 'partially coherent state',

$$\rho(B;z)=\frac{1}{B\,\pi}\int d^2y\,e^{-|y-z|^2/B}\,|y><y|=U(z)\,\rho(B)\,U(z)^{\dagger},\qquad(6.33)$$

for which

$$\begin{cases}\mathrm{Tr}[\rho a]=z\,;\\ \mathrm{Tr}[\rho a^{\dagger}]=z^{*}\,;\\ \mathrm{Tr}[\rho a^{\dagger}a]=z^{*}z+B\,.\end{cases}\qquad(6.34)$$

The set of states (6.31) may easily be generalized to n degrees of freedom,

$$\begin{cases}\rho(p)=\dfrac{1}{\pi^n}\displaystyle\int d^{2n}\mathbf{y}\,p(\mathbf{y})\,|\,\mathbf{y}><\mathbf{y}\,|;\\ p(\mathbf{y})\geq 0\,;\quad \dfrac{1}{\pi^n}\displaystyle\int d^{2n}\mathbf{y}\,p(\mathbf{y})=1.\end{cases}\qquad(6.35)$$

In particular, the partially coherent states obtain, when $p(y)$ is a Gaussian form similar to (6.32),

$$p(\mathbf{y})=\frac{1}{\mathrm{Det}\,\mathbf{B}}\,e^{-(\mathbf{y}^{*}-\mathbf{z}^{*})\cdot(\mathbf{B}^{-1}(\mathbf{y}-\mathbf{z}))},\qquad(6.36)$$

where $\mathbf{B}$ is a positive definite $(n\times n)$-matrix. The density operator corresponding to (6.33) may be written on several forms,

$$\begin{aligned}\rho(\mathbf{B};z)&=U(z)\,\rho(\mathbf{B})\,U(z)^{\dagger}=\\ &=(\mathrm{Det}(\mathbf{1}+\mathbf{B})^{-1})\,e^{-(\mathbf{a}^{\dagger}-\mathbf{z}^{*})\cdot[\ln(\mathbf{1}+\mathbf{B}^{-1})\,(\mathbf{a}-\mathbf{z})]}=\\ &=\frac{1}{\mathrm{Det}\,\mathbf{B}}\cdot\frac{1}{\pi^n}\int d^{2n}\mathbf{y}\,e^{-\mathbf{y}^{*}\cdot(\mathbf{B}^{-1}\mathbf{y})}\,|\,\mathbf{z}+\mathbf{y}><\mathbf{z}+\mathbf{y}\,|=\\ &=\frac{1}{\pi^n}\int d^{2n}\mathbf{y}\,e^{-\mathbf{y}^{*}\cdot\mathbf{y}}\,|\,\mathbf{z}+\mathbf{B}^{1/2}\mathbf{y}><\mathbf{z}+\mathbf{B}^{1/2}\mathbf{y}\,|=\\ &=\exp\!\left(\frac{\partial}{\partial\mathbf{z}}\cdot\left(\mathbf{B}\,\frac{\partial}{\partial\mathbf{z}^{*}}\right)\right)|\,\mathbf{z}><\mathbf{z}\,|\,.\end{aligned}\qquad(6.37)$$

The entropy of (6.37) is

$$S(\mathbf{B}; z) = \mathrm{Tr}[(\mathbf{1} + \mathbf{B}) \ln(\mathbf{1} + \mathbf{B}) - \mathbf{B} \ln \mathbf{B}]. \tag{6.38}$$

Actually (6.36) is the solution of the problem to maximize the entropy under the constraints

$$\begin{cases} \mathrm{Tr}[\rho] = 1; \\ \mathrm{Tr}[\rho \boldsymbol{a}] = \boldsymbol{z}\,; \\ \mathrm{Tr}[\rho \boldsymbol{a}^{\dagger}] = \boldsymbol{z}^{*}\,; \\ \mathrm{Tr}[\rho a_{\mu} a_{\nu}^{\dagger}] = z_{\mu}\, z_{\nu}^{*} + B_{\mu\nu}\,. \end{cases} \tag{6.39}$$

6.2 Stochastic Quantum Dynamics

We consider again a set of n oscillators as described by (6.18). We let them interact with n sources $\boldsymbol{j} = (j_1,\ldots, j_n)$, belonging to the system under study, and a set of external classical sources $\boldsymbol{s}(t) = (s_1(t),\ldots, s_n(t))$, which exhibit a stochastic behaviour. (In general $\boldsymbol{j}$ is non-Hermitean and $\boldsymbol{s}(t)$ is complex-valued.) The external source is assume to fluctuate around a vanishing mean

$$\begin{cases} < \boldsymbol{s}(t) > = 0; \\ < s_k(t_1)\, s_l(t_2) > = < s_k^{*}(t_1)\, s_l^{*}(t_2) > = 0\,, \quad \text{all } t_1, t_2\,; \\ < s_k(t)\, s_l^{*}(t') > = b_{kl}(t)\, \delta(t - t')\,; \quad b_{kl}^{*}(t) = b_{lk}(t), \end{cases} \tag{6.40}$$

where $\mathbf{b}(t)$ is a positive definite matrix. Thus $\int dt'\, \boldsymbol{s}(t')$ describes a Brownian motion in the $2N$-dimensional space.

We shall use this model to determine the time-development of the quantum system, due to the influence of the external source $\boldsymbol{s}(t)$. The system may be described by the density operator ρ in the interaction picture, where

$$a_k(t) = e^{i\omega_k t}\, a_k(0). \tag{6.41}$$

Then

$$i\,\frac{d}{dt}\,\rho(t) = [H_I(t) + H_X(t), \rho(t)], \tag{6.42}$$

where

$$\begin{cases} H_I(t) = i\,(\boldsymbol{a}^\dagger(t)\cdot\boldsymbol{j}(t) - \boldsymbol{j}^\dagger(t)\cdot\boldsymbol{a}(t)); \quad \boldsymbol{j}(t) = e^{-iH_j t}\,\boldsymbol{j}(0)\,e^{iH_j t}; \\ H_X(t) = i\,(\boldsymbol{a}^\dagger(t)\cdot\boldsymbol{s}(t) - \boldsymbol{s}^*(t)\cdot\boldsymbol{a}(t)). \end{cases} \tag{6.43}$$

Here H_j is the free Hamiltonian of the source field.

Let now $\overline{\rho}(t)$ be the density operator obtained from averaging $\rho(t)$ over $\boldsymbol{s}(t')$. We are looking for the equation of motion for $\overline{\rho}(t)$. Then for the change during a small time interval Δt we find

$$\frac{1}{\Delta t}\,(\overline{\rho}(t+\Delta t) - \overline{\rho}(t)) =$$

$$= \frac{1}{i}\,[H_I(t), \overline{\rho}(t)] + \frac{1}{i\,\Delta t}\left[<\int_t^{t+\Delta t} dt'\, H_X(t')>,\ \overline{\rho}(t)\right] -$$

$$\frac{1}{2\,\Delta t} < \left[\int_t^{t+\Delta t} dt'\, H_X(t'), \left[\int_t^{t+\Delta t} dt''\ H_X(t''), \overline{\rho}(t)\right]\right] > + O(\Delta t) \tag{6.44}$$

On the right side '< >' denotes an averaging over $\boldsymbol{s}(t')$, $t < t' < t + \Delta t$. Using (6.43), (6.41) and (6.40), we get in the limit $\Delta t \to 0$, the Liouville equation

$$i\,\frac{d}{dt}\,\overline{\rho}(t) = \overline{\mathcal{H}}(t)\ \overline{\rho}(t). \tag{6.45}$$

The Liouville operator $\overline{\mathcal{H}}(t)$ is defined as follows (χ = arbitrary density operator)

$$\bar{\mathcal{H}}(t) = \mathcal{H}_{\rm I}(t) + \bar{\mathcal{H}}_{\rm X}(t), \tag{6.46}$$

where [Lindblad 1985]

$$\mathcal{H}_{\rm I}(t)\,\chi = [H_{\rm I}(t), \chi], \tag{6.47}$$

and

$$\bar{\mathcal{H}}_{\rm X}(t)\,\chi = \frac{1}{i}\sum_{k,l=1}^{n} b_{kl}(t)\,{\rm e}^{i(\omega_l - \omega_k)t}\,[a_k^\dagger, [a_l, \chi]]. \tag{6.48}$$

Let $B^{(r)}(t)$ be the eigenvalues of $b_{kl}(t)\exp[i(\omega_l - \omega_k)t]$, and $U(t)$ the diagonalizing unitary transformation. Then

$$\begin{cases} b_{kl}(t)\,{\rm e}^{i(\omega_l - \omega_k)t} = \sum_r U_{kr}(t)\,B^{(r)}(t)\,U_{lr}^*(t); \\ B^{(r)}(t) \geq 0\,; \quad \sum_r B^{(r)}(t) > 0; \\ \sum_r U_{kr}(t)\,U_{lr}^*(t) = \delta_{kl}\,. \end{cases} \tag{6.49}$$

Then introducing the transformed oscillators

$$\begin{cases} A_r(t) = \sum_k a_k U_{kr}^*(t); \\ [A_r(t), A_s^\dagger(t)] = \delta_{rs}\,, \end{cases} \tag{6.50}$$

we can rewrite (6.48) as

$$\bar{\mathcal{H}}_{\rm X}(t)\,\chi = \frac{1}{i}\sum_r B^{(r)}(t)\,[A_r^\dagger(t), [A_r(t), \chi]]. \tag{6.51}$$

Clearly probability is automatically conserved by (6.47) and (6.51),

$$i\,\frac{{\rm d}}{{\rm d}t}\,{\rm Tr}(\bar{\rho}(t)) = {\rm Tr}(\bar{\mathcal{H}}(t)\,\bar{\rho}(t)) = 0. \tag{6.52}$$

To see the effect of the equation of motion (6.45), we may study the diagonal elements of $\overline{\rho}$,

$$p_\alpha(t) = \overline{\rho}_{\alpha\alpha}(t). \tag{6.53}$$

In the representation where $\overline{\rho}$ is diagonalized at time t, we have (α, β denote quantum states)

$$\frac{\mathrm{d}}{\mathrm{d}t}\, p_\alpha(t) = \sum_\beta M_{\alpha\beta}(t)\,(p_\beta(t) - p_\alpha(t)), \tag{6.54}$$

where

$$M_{\alpha\beta}(t) = \sum_r B^{(r)}(t)\,(\,|(A_r(t))_{\alpha\beta}|^2 + |(A_r(t))_{\beta\alpha}|^2) = M_{\beta\alpha}(t) \geq 0. \tag{6.55}$$

For the entropy production we find in the same representation

$$\dot{S}(t) = -\frac{\mathrm{d}}{\mathrm{d}t}\,\mathrm{Tr}[\,\overline{\rho}(t)\ln\overline{\rho}(t)] = -\,\mathrm{Tr}\left[\frac{\mathrm{d}\,\overline{\rho}(t)}{\mathrm{d}t}\ln\overline{\rho}(t)\right] =$$

$$= \frac{1}{2}\sum_{\alpha,\beta} M_{\alpha\beta}(t)\,(p_\alpha(t) - p_\beta(t))\ln\frac{p_\alpha(t)}{p_\beta(t)} \geq 0. \tag{6.56}$$

The solution of (6.45) for a special case will be given below in Section 6.3.

The Hermitean conjugate $\mathcal{A}^\dagger$ of a "super-operator" $\mathcal{A}$ in operator space is defined through

$$\mathrm{Tr}[\chi_1\,(\mathcal{A}^\dagger\chi_2)] = \mathrm{Tr}[(\mathcal{A}\chi_1)^\dagger\chi_2]. \tag{6.57}$$

Then we see that $\overline{\mathcal{H}}(t)$ is composed of a Hermitean part and an anti-Hermitean part,

$$\begin{cases} \mathcal{H}_\mathrm{I}^\dagger(t) = \mathcal{H}_\mathrm{I}(t), \\ \overline{\mathcal{H}}_\mathrm{X}^\dagger(t) = -\,\overline{\mathcal{H}}_\mathrm{X}(t). \end{cases} \tag{6.58}$$

Thus the effect of the external stochastic current is described by an anti-Hermitean term in the Liouville operator.

Under time-reversal T,

$$\begin{cases} \boldsymbol{a}(t) \to \boldsymbol{a}^{(T)}(-t) = \boldsymbol{a}(t), \\ \boldsymbol{j}(t) \to \boldsymbol{j}^{(T)}(-t) = -\boldsymbol{j}(t), \\ \boldsymbol{s}(t) \to \boldsymbol{s}^{(T)}(-t) = -\boldsymbol{s}(t). \end{cases} \tag{6.59}$$

Furthermore, $\mathcal{H}_{\mathrm{I}}(t)$ is odd and $\overline{\mathcal{H}}_{\mathrm{X}}(t)$ is even,

$$\begin{cases} \mathcal{H}_{\mathrm{I}}^{(T)}(-t) = -\mathcal{H}_{\mathrm{I}}(t), \\ \overline{\mathcal{H}}_{\mathrm{X}}^{(T)}(-t) = \overline{\mathcal{H}}_{\mathrm{X}}(t). \end{cases} \tag{6.60}$$

This is in agreement with Prigogine's microscopic theory of irreversible processes [Prigogine 1980, Chapter 8]. Under the Prigogine 'star' operation

$$\mathcal{A}^{*} = \mathcal{A}^{(T)\dagger}, \quad t \to -t, \tag{6.61}$$

due to (6.58) and (6.60), $\mathcal{H}_{\mathrm{I}}(t)$ and $\overline{\mathcal{H}}_{\mathrm{X}}(t)$ are odd, and hence also the total Hamiltonian super-operator $\overline{\mathcal{H}}(t)$,

$$\begin{cases} \mathcal{H}_{\mathrm{I}}^{*}(-t) = -\mathcal{H}_{\mathrm{I}}(t), \\ \overline{\mathcal{H}}_{\mathrm{X}}^{*}(-t) = \overline{\mathcal{H}}_{\mathrm{X}}(t), \\ \overline{\mathcal{H}}^{*}(-t) = \overline{\mathcal{H}}(t). \end{cases} \tag{6.62}$$

Thus, (6.45) is invariant under the star operation.

In the space of operators the vectors corresponding to coherent states in the Hilbert space of states are of the type

$$\chi = | \boldsymbol{z} \rangle \langle \boldsymbol{x} |, \tag{6.63}$$

where $|z>$ and $|x>$ are coherent states, as defined in (6.24),

$$\begin{cases} a\,|\,z> = z\,|\,z> \\ <x\,|\,a^{\dagger} = x^{*}<x\,| \end{cases} \qquad \begin{cases} <z\,|\,z> = 1 \\ <x\,|\,x> = 1 \end{cases} \tag{6.64}$$

Then χ is also normalized,

$$\mathrm{Tr}(\chi^{\dagger}\chi) = <x\,|\,x><z\,|\,z> = 1. \tag{6.65}$$

Like the coherent states in Hilbert space, vectors of the kind (6.63) form an overcomplete basis in operator space.

The "expectation value" of the Liouville operator $i\ \overline{\mathcal{H}}_{\mathrm{X}}$ for the vector (6.63) is

$$\mathrm{Tr}[\chi^{\dagger}(i\ \mathcal{H}_{\mathrm{X}}(t)\chi)] = \sum_{r} B^{(r)}(t)\ (1 + |Z_r(t) - X_r(t)|^2) > 0, \tag{6.66}$$

where

$$\begin{cases} X_r(t) = \sum\limits_{k} x_k\ U^{*}_{kr}(t), \\ Z_r(t) = \sum\limits_{k} z_k\ U^{*}_{kr}(t). \end{cases} \tag{6.67}$$

This shows that $i\ \overline{\mathcal{H}}_{\mathrm{X}}$ is positive definite, i.e., $\overline{\mathcal{H}}_{\mathrm{X}}$ acts as an absorptive potential.

6.3 Phase-Space Geometry

To study geometrical aspects of information in quantum mechanics we chose again the system of n oscillators. The space of interest here is the phase space, and as a basis we shall use the coherent states (6.24), each of which corresponds to a point in phase space. The Hilbert space metric induces a metric in phase space,

$$|z - \boldsymbol{x}|^2 = \ln \frac{1}{|\langle z | \boldsymbol{x} \rangle|^2} . \tag{6.68}$$

The concept of resolution is still useful. If we have a 'true' macrostate ρ, then for a resolution r the corresponding 'observed' macrostate is (see (3.6))

$$\rho(r) = \pi^{-n} \int d^{2n}z \, e^{-z^* \cdot z} \, U(r\,z)\rho U(-r\,z) = e^{-r^2 a^\dagger_\mu a_\mu} \rho , \tag{6.69}$$

where the operator a in operator space is defined by (we use here the convention of summation over repeated vector indices)

$$a_\mu \rho = [a_\mu, \rho]. \tag{6.70}$$

The smeared density operator (6.69) satisfies a diffusion equation

$$\left(\frac{d}{dr} + 2r\, a^\dagger_\mu \, a_\mu\right) \rho(r) = \frac{d\rho(r)}{dr} + 2r \, [a^\dagger_\mu, [a_\mu, \rho]] = 0. \tag{6.71}$$

This immediatly gives us a possibility of solving the equation of motion (6.45) in a special case. We choose a set of oscillators decoupled from the source field and interacting only with the stochastic source, which is assumed to have constant properties. We thus put $H_{\mathrm{I}} = 0$ in (6.47) and (6.46) and we assume for $b_{kl}(t)$, appearing in (6.40) and (6.48),

$$b_{kl}(t) = \hat{b}_{kl} \, e^{i(\omega_k - \omega_l)t} \, ; \quad b^*_{kl} = \hat{b}_{lk} \, , \tag{6.72}$$

where $\hat{b}_{kl}$ is constant in time. Then (6.46) implies

$$\frac{d}{dt} \, \bar{\rho}(t) = - \, a^\dagger_\mu \hat{b}_{\mu\nu} \, a_\nu \, \bar{\rho}(t), \tag{6.73}$$

with the result

$$\bar{\rho}(t) = e^{-\frac{1}{2} a_\mu^\dagger \hat{b}_{\mu\nu} a_\nu (t-t_o)} \bar{\rho}(t_o) ; \quad t > t_o , \tag{6.74}$$

or, because of (6.70), more explicitly

$$\bar{\rho}(t) = \frac{1}{\text{Det}\,\hat{\mathbf{b}}} \frac{1}{(2\pi (t-t_o))^n} \int d^{2n} z \, e^{-\frac{1}{2(t-t_o)} z^* \cdot (\hat{\mathbf{b}}^{-1} z)} U(z) \bar{\rho}(t_o) U(-z). \tag{6.75}$$

For the contrast (6.10), we shall now make a double decomposition with respect to position in phase space and resolution r, related to structure size in phase space. The assumption of an isotropic resolution involves a choice of scale in coordinate space (Re z) and momentum space (Im z).

If ρ_o is so wide that it is practically constant where ρ has support, and practically uneffected by the smearing (6.69), then

$$K[\rho_o; \rho] = -\int_0^\infty dr \frac{d}{dr} K[\rho_o; \rho(r)] = -\int_0^\infty dr \, \text{Tr}\left[\frac{d\rho(r)}{dr} \ln \rho(r)\right] =$$

$$= \int_{-\infty}^{\infty} d\left(\ln \frac{1}{r}\right) 2 r^2 \, \text{Tr}\left([a_\mu^\dagger, [a_\mu, \rho(r)]] \ln \rho(r)\right). \tag{6.76}$$

We can rewrite the trace here as

$$\text{Tr}([a_\mu, \rho(r)] [\ln \rho(r), a_\mu^\dagger]) = \text{Tr}\left(\rho(r)\left[a_\mu, \frac{1}{\rho(r)}\right]\rho(r)\left[\ln \frac{1}{\rho(r)}, a_\mu^\dagger\right]\right) =$$

$$= \sum_{j,l=0}^{\infty} \frac{1}{(j+l+1)!} \text{Tr}\left(\rho(r)\left(\ln \frac{1}{\rho(r)}\right)^j \left[a_\mu, \ln \frac{1}{\rho(r)}\right]\left(\ln \frac{1}{\rho(r)}\right)^l \rho(r)\left[\ln \frac{1}{\rho(r)}, a_\mu^\dagger\right]\right) =$$

$$= \sum_{j,l=0}^{\infty} \text{Tr}(\boldsymbol{h}_{jl}(r) \cdot \boldsymbol{h}_{jl}(r)^\dagger), \tag{6.77}$$

where

$$\boldsymbol{h}_{jl}(r) = \sqrt{\rho(r)} \left(\ln \frac{1}{\rho(r)}\right)^{j/2} \left[\boldsymbol{a}, \ln \frac{1}{\rho(r)}\right]\left(\ln \frac{1}{\rho(r)}\right)^{l/2} \sqrt{\rho(r)} \, . \tag{6.78}$$

Thus

$$K[\rho_o;\rho] = \int_{-\infty}^{\infty} \mathrm{d}\left(\ln\frac{1}{r}\right)\frac{1}{\pi^n}\int \mathrm{d}^{2n}z < z\,|\,k(r)\,|\,z >, \tag{6.79}$$

where we have introduced the (positive definite) information density operator,

$$k(r) = 2\,r^2 \sum_{j,\,l=0}^{\infty} \boldsymbol{h}_{jl}(r)\cdot\boldsymbol{h}_{jl}(r)^{\dagger}. \tag{6.80}$$

In (6.79) we can now interpret $\pi^{-n}<z|k(r)|z>$ as the amount of information per unit of $\ln(1/r)$ and unit volume in phase space.

We note that for a density operator of the type (6.35),

$$\rho = \pi^{-n}\int \mathrm{d}^{2n}z\, p(z)\,|\,z> <z\,|\,, \tag{6.81}$$

the probability density corresponding to the resolution r is

$$p(r;z) = \pi^{-n}\int \mathrm{d}^{2n}\boldsymbol{y}\,\mathrm{e}^{-\boldsymbol{y}^*\cdot\boldsymbol{y}}\,p(z+r\,\boldsymbol{y}), \tag{6.82}$$

and the corresponding density operator is

$$\rho(r) = \pi^{-n}\int \mathrm{d}^{2n}z\, p(r;z)\,|\,z> <z\,|\,. \tag{6.83}$$

Instead of (6.79) we can now use the form

$$K[\rho_o;\rho] = \int_{-\infty}^{\infty} \mathrm{d}\left(\ln\frac{1}{r}\right)\pi^{-n}\int \mathrm{d}^{2n}z\, p(r;z) < z\,|\,(2n-d(r))\,|\,z >, \tag{6.84}$$

where $d(r)$ is the local dimension operator for the resolution r,

$$d(r) = 2n - 2r^2 \sum_{j,\,l=0}^{\infty} \frac{1}{\sqrt{\rho(r)}}\,\boldsymbol{h}_{jl}(r)\cdot\boldsymbol{h}_{jl}(r)^{\dagger}\,\frac{1}{\sqrt{\rho(r)}}\,. \tag{6.85}$$

6.4 Quantum Computers

The idea of a Hamiltonian quantum system carrying out a computation was formulated and discussed by Benioff [1982]. A detailed theory was worked out by Feynman [1985]. If it is reversible, its time evolution is described by a unitary evolution operator

$$U(t) = e^{-iHt}, \tag{6.86}$$

where H is a Hermitean Hamiltonian. Conveniently, it may be chosen as a sum of terms, each corresponding to a computational step and its inverse

$$H = \sum_{i=1}^{n} (A_i \otimes |i>_{c\,c}<i-1| + A_i^{\dagger} \otimes |i-1>_{c\,c}<i|); \quad A_i A_i^{\dagger} = 1, \tag{6.87}$$

where $|i>_c$ $(i = 0, 1, \ldots, n)$ is an orthogonal basis for the 'cursor', which keeps track of the progress of computation. Cursor index 0 represents input and n represents output. The computer is assumed to have N binary sites. Its state is characterized by the state of each site. The ith site has the basis $|0>_i$ and $|1>_i$. The binary string $a_1 a_2 \ldots a_N$ then corresponds to the state

$$|a_1>_1 |a_2>_2 \ldots |a_N>_N; \qquad a_i = 0, 1; \quad i = 1, \ldots, N. \tag{6.88}$$

We assume that the 2^N different string states (6.88) form a complete orthonormal basis for the states of the binary sites.

In an expansion of (6.86), the terms that are effective in leading an input

$$|a_1> \ldots |a_N> |0>_c \tag{6.89}$$

through the computer all add up to

$$(-i)^n f(t)\, A_n \ldots A_2 A_1 \otimes |n>_{c\,c}<0|; \quad f(t)^* = f(t). \tag{6.90}$$

This follows from the properties of the cursor factors and the unitarity of the A_i operators. Corresponding to (6.90) there is a term in the expansion of (6.86) undoing the computation

$$i^n f(t)\, A_1^\dagger A_2^\dagger \ldots A_n^\dagger \otimes | 0 >_{c\ c}< n | . \tag{6.91}$$

To speed up the computation it should be possible to give the signal travelling through the computer a momentum in the direction of increasing cursor index. The corresponding kinetic energy is present in the input, and may be extracted from the output. To build up a suitable wave, and to extract the output energy, one may need a lattice in front of the computer, for the incoming wave to travel along, and similarly behind the computer for the output wave.

All this is very similar in spirit to the classical billiard ball computer (Section 4.5). As Zurek [1984] has shown, due to the unitarity of $U(t)$, an error in the input,

$$\sqrt{1-\alpha^2}\, | \phi_{\text{in}} > + \alpha\, | \psi > ; \qquad < \psi | \phi_{\text{in}} > = 0, \tag{6.92}$$

instead of $|\phi_{\text{in}}>$, leads to an error of the same size at any time:

$$\sqrt{1-\alpha^2}\, U(t)\, | \phi_{\text{in}} > + \alpha\, U(t)\, | \psi > \tag{6.93}$$

represents $U(t)\, |\phi_{\text{in}}>$ just as well as (6.92) represents $|\phi_{\text{in}}>$.

A hardware error modifying the Hamiltonian H into

$$H' = H + h\,, \tag{6.94}$$

leads during a time t to a state $e^{-i(H+h)t}|\phi_{\text{in}}>$ to be compared with $e^{-iHt}|\phi_{\text{in}}>$. A good measure of this deviation is [Zurek 1984]

$$1 - |< \phi_{\text{in}} | e^{iHt}\, e^{-i(H+h)t} | \phi_{\text{in}} >|^2 \approx < (h - < h >_{\text{in}})^2 >_{\text{in}} t^2\,, \tag{6.95}$$

where $< >_{\text{in}}$ denotes expectation value for the state $|\phi_{\text{in}}>$. In (6.95) we have assumed that $<[H, h]>_{\text{in}}$ is negligible. If we set t equal to the computation time, we see that by increasing the momentum of the wave-packet one can make the derivation time t and hence also the deviation (6.95) very small.

The quantum computer is in this respect similar to a classical ballistic computer, but as in the classical case, ideal (non-thermal) conditions are quite unlikely. Therefore we should look at a Brownian quantum computer. It would obey an equation of motion of the type (6.45). But as we have seen, this would lead to dissipative processes very similar to the classical case. A Brownian quantum computer would dissipate just like a classical Brownian computer discussed in Section 4.5.1. This is also the conclusion drawn by Feynman [1985].

6.5 Open Quantum-Mechanical Systems

When George Gamow wanted to explain quantum phenomena by showing what they would look like on a macroscopic scale, he sent out his alter ego, Mr Tompkins, on a tiger-hunting tour [Gamow 1965]. Mr Tompkins does not know where to aim in the "large pack of fuzzy-looking tigers".

In physical experiments and physics text-books we have learnt that quantum phenomena on the atomic and subatomic levels are very similar to what Gamow describes. Microscopic states may be linearly superposed into new states which are different from but still somehow contain the features of the states from which they are superposed. On the macroscopic level however, this is not the case. Classically, different states do not occur together in a superposition. Tigers are not fuzzy; they are well localized. The classical every-day world we experience is—in this respect—unambiguous.

Somewhere along the way between the atomic level and the macroscopic level there is thus a bifurcation mechanism which involves a choice, such that the resulting superposition only contains microstates which are similar on the macroscopic level.

Related to Mr Tompkins' tiger, there is another cat, Schrödinger's cat, which was introduced to function as the pointer of a measuring instrument. The microsystem to be subject to measurement is in a superpostion of two different eigenstates of the observable to be measured. The measuring device is so constructed that if one eigenvalue results, nothing happens to the cat, but if the other eigenvalue obtains, the cat is killed. There have been speculations about quantum-mechanical superpostions of one state with the cat alive and another one with the cat dead,

$\frac{1}{\sqrt{2}}$ (|  > + |  >).

Much has been written about this Gedanken-experiment. According to the above reasoning, such superpositions do not occur; the wave-function is reduced, before the cat is affected.

The point of view taken here is that the bifurcations taking place between micro and macro are *not* due to

i) the intervention of the human mind,
ii) the splitting of the universe into parallel universes,
iii) quantum jumps,
iv) hidden variables, not present in the (incomplete) quantum-mechanical description,
v) an inherent, unavoidable, incompleteness in the theory of quantum mechanics itself.

We shall see that the bifurcations—which make the quantum-mechanical world appear classical on a macroscopic level—need no such drastic ontological or epistemological assumptions to be understood. These bifurcations arise simply from the laws of large numbers, from the statistical mechanics of open quantum-mechanical systems. A single measuring process can be modelled in a way very similar to the way in which a single fluctuation-triggered classical self-organization process may be modelled.

It then appears very reasonable to accept a simple quantum-mechanical ontology, i.e., an ontology which ascribes reality to the quantum-mechanical state.

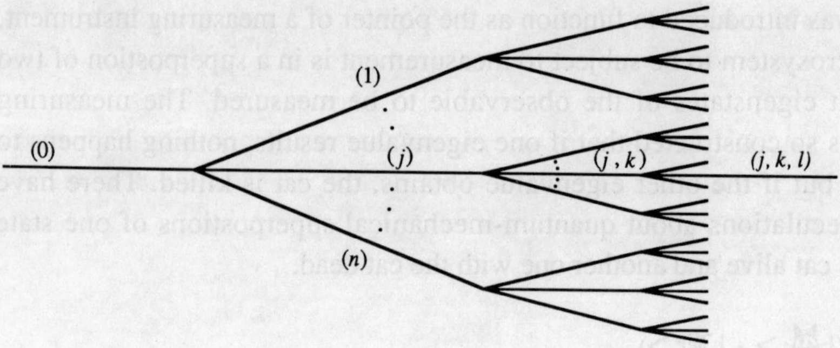

Fig. 6.1. Quantum-mechanical branching.

We construct a class C of open quantum-mechanical systems (OQS) in the following way. The system is originally in the state $|0>$, the "trunk" state (Fig. 6.1). Transitions are allowed from $|0>$ into the branch states $|j, 0>$ ($j = 1,\ldots, n$) and from these into sub-branch states of increasing order $|j, 1; k_1>, \ldots, |j, M; k_1 \ldots k_M>, \ldots, |j, L; k_1 \ldots k_L>$. For simplicity we assume that—except at the first branching—the coupling strength is the same everywhere in the tree. For the branch j, we assume $k_1 = 1,\ldots, m_1^{(j)}$; $\ldots$; $k_L = 1,\ldots, m_L^{(j)}$, where $m_1^{(j)},\ldots, m_L^{(j)}$ are stochastic multiplicities to be discussed below. At the Lth subbranch level we assume irreversible absorption—uniform among the branches.

The interaction Hamiltonian describing this situation is (h.c. = Hermitean conjugate)

$$H_L = \sum_{j=1}^{n} \Big[g_0 \sqrt{\alpha_j} \, (| j, 0 > < 0 | + \text{h.c.})$$

$$+ g \sum_{M=1}^{L} \sum_{k_1=1}^{m_1^{(j)}} \cdots \sum_{k_M=1}^{m_M^{(j)}} (| j, M; k_1 \ldots k_M > < j, M-1; k_1 \ldots k_{M-1} | + \text{h.c.}) \Big]$$

$$- iu \sum_{j=1}^{n} \sum_{k_1=1}^{m_1^{(j)}} \cdots \sum_{k_L=1}^{m_L^{(j)}} | j, L; k_1 \ldots k_L > < j, L; k_1 \ldots k_L | , \qquad (6.96)$$

which in the limit of infinite r is

$$f(p,\infty)=p_o\delta(p-1)+(1-p_o)\delta(p). \tag{6.114}$$

In Fig. 6.2 $f(p, r)$ is plotted as a function of p for some values on r, with $p_o=0.7$. In the case of general n, starting from $f(\boldsymbol{p}, 0) = \delta(\boldsymbol{p} - \alpha)$, one ends up with

$$f(\boldsymbol{p},\infty)=\sum_{j=1}^{n}\alpha_j\prod_{k=1}^{n}\delta(p_k-\delta_{kj}). \tag{6.115}$$

The approach to this limit is very fast. For instance, the ensemble average of information

$$K(r)=\int_{\boldsymbol{p}>0}\mathrm{d}^n\boldsymbol{p}\,\delta\Big(\sum_j p_j-1\Big)f(\boldsymbol{p},r)\sum_{j=1}^{n}p_j\ln\frac{p_j}{\alpha_j}, \tag{6.116}$$

which obeys an equation of motion

$$\frac{\mathrm{d}}{\mathrm{d}r}K(r)=\int_{\boldsymbol{p}>0}\mathrm{d}^n\boldsymbol{p}\,\delta\Big(\sum_j p_j-1\Big)f(\boldsymbol{p},r)\sum_{j=1}^{n}p_j(1-p_j)\geq 0, \tag{6.117}$$

approaches its asymptotic value

$$K(\infty)=\sum_{j=1}^{n}\alpha_j\ln\frac{1}{\alpha_j} \tag{6.118}$$

as

$$K(\infty)-K(r)<4re^{-4r}\max_j\left(\frac{p_j}{1-p_j}\right). \tag{6.119}$$

Because of (6.101) this means that only a rather moderate value of $\langle N\rangle$ is needed to reach the asymptotic region.

The individual tree is characterized by a random walk ending up in one of the corners of $X^{(n-1)}$. Fig. 6.3 shows two examples of random walks for $n = 2$.

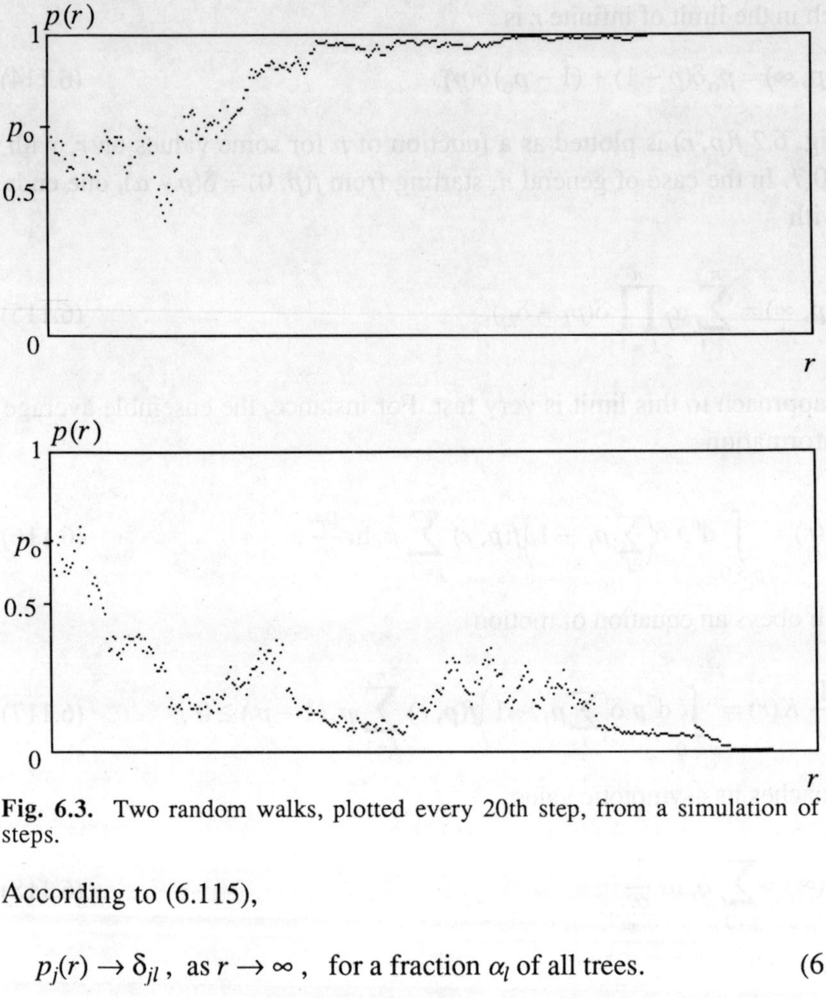

Fig. 6.3. Two random walks, plotted every 20th step, from a simulation of 7600 steps.

According to (6.115),

$$p_j(r) \to \delta_{jl}\,, \quad \text{as } r \to \infty\,, \quad \text{for a fraction } \alpha_l \text{ of all trees.} \tag{6.120}$$

Thus the statistics for the class C of OQS leads to a definite choice. If for large r, i.e., large L, sub-branches with different branch indices j correspond to macroscopically different states, then there will be no superposition of such states. There is no superposition of states with a dead Schrödinger's cat and states with the same cat alive!

It should be emphasized that the motion (6.106) in the probability simplex depends crucially on (6.104), which rests on quantum-mechanical, not classical, dynamics. In fact, (6.104) results from unitary dynamics (apart from absorption) of the Hamiltonian H_L. The nonlinear and random character of (6.106) comes from mapping the result from one H_L on the results from the following ones.

In the unitary evolution, the state $|0>$ is a necessary link between the different branches.

6.5.1 *Theory of quantum measurement*

The present theory of OQS can be immediately used to construct a model of quantum measurement [Wheeler and Zurek 1983]. Let μ denote 'micro-system' (to-be-measured) and M 'macrosystem' (measuring apparatus), and let

$$|\psi>_\mu = \sum_{j=1}^{n} \sqrt{\alpha_j}\, |j>_\mu \tag{6.121}$$

be the original state of the microsystem in a basis where the operator T corresponding to the measurement observable is diagonal,

$$T\,|j>_\mu = t_j\,|j>_\mu \qquad (t_j = t_k \text{ only for } j = k). \tag{6.122}$$

Further let $|0>_{\mathrm{M}}$ denote a state of readiness in M, and let $|j, L; k_1...k_L>_{\mathrm{M}}$ denote states which with increasing L approach a macroscopic recording of j. With

$$\begin{cases} |0> = |\psi>_\mu \otimes |0>_{\mathrm{M}}; \\ |j, 0> = |j>_\mu \otimes |j, 0>_{\mathrm{M}}; \\ |j, M; k_1\ldots k_M> = |j>_\mu \otimes |j, M; k_1\ldots k_M>_{\mathrm{M}} \qquad (M = 1,\ldots, L), \end{cases} \tag{6.123}$$

The Hamiltonian (6.96) is equivalent to

$$H_L = g_0 P \sum_{j=1}^{n} |j>_{\mu\ \mu}<j| \otimes (|j,0>_{M\ M}<0| + \text{h.c.})\, P +$$

$$+ 1_\mu \otimes \sum_{j=1}^{n} \Bigg[g \sum_{M=1}^{L} \sum_{k_1=1}^{m_1^{(j)}} \ldots \sum_{k_M=1}^{m_M^{(j)}} (|j, M; k_1 \ldots k_M> <j, M-1; k_1 \ldots k_{M-1}| + \text{h.c.})$$

$$- iu \sum_{k_1=1}^{m_1^{(j)}} \ldots \sum_{k_L=1}^{m_L^{(j)}} |j, L; k_1 \ldots k_L> <j, L; k_1 \ldots k_L| \Bigg] . \qquad (6.124)$$

Here the interaction between μ and M takes place only through the first term. The projector

$$P = |\psi>_{\mu\ \mu}<\psi| \otimes |0>_{M\ M}<0| + 1_\mu \otimes (1_M - |0>_{M\ M}<0|) \qquad (6.125)$$

guarantees that M does not return to its state of readiness without restoring μ in its original state $|\psi>_\mu$. Thus, the first term in (6.124) is a copying/de-copying operator, where P ties the readiness state of M to the preparation of μ in the state $|\psi>_\mu$. This is an essential feature in this model of quantum measurement. Like in the quantum computer Hamiltonian, to each step forward in (6.124), there is also a term for the reverse step. The state $|0>$ as given in (6.123) is the trunk of the tree of quantum-machanical branching (Fig. 6.1).

6.5.2 *A truncated model for the 'choice'*

We shall construct a model for describing the action of M on μ in the case when channel l is chosen, i.e., when

$$\frac{\alpha_j}{\alpha_l} \prod_{L=1}^{N} \frac{m_L^{(j)}}{m_L^{(l)}} << 1\,, \quad \text{all } j \neq l\,. \qquad (6.126)$$

To simplify the description we use only one state $|j>_M$

channel (corresponding to $|j, 0\rangle_M$ above), and we consider absorption as induced from $|l\rangle_M$; the absorption from the other channels vanish. We thus contract the whole chain of interactions (including absorption) into one term describing absorption in the chosen channel l. Moreover we put $g_0=1$, thereby fixing the time scale.

Our Hamiltonian is then

$$H = \sum_{j=1}^{n} \sqrt{\alpha_j}\, (|j\rangle_\mu\, {}_\mu\langle \psi| \otimes |j\rangle_M\, {}_M\langle 0| + \text{h.c.})$$
$$- i\gamma 1_\mu \otimes |l\rangle_M\, {}_M\langle l|\,, \qquad (6.127)$$

where $\gamma > 0$. At the time $t = 0$ the intial state is

$$|0\rangle = |\psi\rangle_\mu \otimes |0\rangle_M\,; \quad |\psi\rangle_\mu = \sum_{j=1}^{n} \sqrt{\alpha_j}\, |j\rangle_\mu\,; \quad \sum_{j=1}^{n} \alpha_j = 1\,. \qquad (6.128)$$

It is convenient to introduce the notation

$$\begin{cases} |l\rangle = |l\rangle_\mu \otimes |l\rangle_M\,; \\ |x\rangle = \dfrac{1}{\sqrt{1-\alpha_l}} \displaystyle\sum_{j \neq l} \sqrt{\alpha_j}\, |j\rangle_\mu \otimes |j\rangle_M\,. \end{cases} \qquad (6.129)$$

The action of H on $|0\rangle$ is then restricted to the space spanned by $|0\rangle$, $|l\rangle$ and $|x\rangle$,

$$e^{-iHt}|0\rangle = \varphi_0(t)\,|0\rangle + \varphi_l(t)\,|l\rangle + \varphi_x(t)\,|x\rangle\,. \qquad (6.130)$$

Due to absorption this state is not normalized for $t > 0$, but

$$p_{\text{abs}}(t) = 1 - |\varphi_0(t)|^2 - |\varphi_l(t)|^2 - |\varphi_x(t)|^2 \geq 0\,, \qquad (6.131)$$

with equality only for $t = 0$.

The state vector satisfies a Schrödinger equation

$$\left[i\frac{\mathrm{d}}{\mathrm{d}t} + \begin{pmatrix} 0 & \sqrt{\alpha_l} & \sqrt{1-\alpha_l} \\ \sqrt{\alpha_l} & -i\gamma & 0 \\ \sqrt{1-\alpha_l} & 0 & 0 \end{pmatrix} \right] \begin{pmatrix} \varphi_o(t) \\ \varphi_l(t) \\ \varphi_x(t) \end{pmatrix} = 0\,, \tag{6.132}$$

with initial condition

$$\begin{pmatrix} \varphi_o(0) \\ \varphi_l(0) \\ \varphi_x(0) \end{pmatrix} = \begin{pmatrix} 1 \\ 0 \\ 0 \end{pmatrix}, \tag{6.133}$$

For the absorption probability (6.131) we have

$$\begin{cases} \dot{p}_{\mathrm{abs}}(t) = 2\gamma |\varphi_l(t)|^2\,, \\ p_{\mathrm{abs}}(0) = 0\,. \end{cases} \tag{6.134}$$

The Schrödinger equation may be transformed into the differential equation

$$\dddot{\varphi}_o(t) + \gamma\,\ddot{\varphi}_o(t) + \dot{\varphi}_o(t) + \gamma(1-\alpha_l)\,\varphi_o(t) = 0\,, \tag{6.135}$$

with initial conditions

$$\begin{cases} \varphi_o(0) = 1\,, \\ \dot{\varphi}_o(0) = 0\,, \\ \ddot{\varphi}_o(0) = -1\,, \end{cases} \tag{6.136}$$

from which we deduce that $\varphi_o(t)$ is real. The other components may then be determined from $\varphi_o(t)$,

$$\begin{cases} \varphi_l(t) = \dfrac{i}{\gamma\sqrt{\alpha_l}}\,(\ddot{\varphi}_o(t) + \varphi_o(t)) = -\,\varphi_l(t)^*\,; \\ \varphi_x(t) = \dfrac{i}{\gamma\sqrt{1-\alpha_l}}\,(\ddot{\varphi}_o(t) + \gamma\dot{\varphi}_o(t) + \varphi_o(t)) = -\,\varphi_x(t)^*\,. \end{cases} \tag{6.137}$$

The density operator for μ is $(j, j' \neq l)$

$$\rho_\mu(t) = \mathrm{Tr_M}\,(e^{-iHt}\,|\,0><0\,|\,e^{iHt}) + p_{\mathrm{abs}}(t)\,|\,l>_{\mu\ \mu}<l\,| =$$

$$= (-\,\varphi_l(t)^2 + \alpha_l\,\varphi_o(t)^2 + p_{\mathrm{abs}}(t))\,|\,l>_{\mu\ \mu}<l\,| +$$

$$+ (-\,\varphi_x(t)^2 + (1-\alpha_l)\,\varphi_o(t)^2)\,|\,x>_{\mu\ \mu}<x\,| +$$

$$+ \sqrt{\alpha_l\,(1-\alpha_l)}\;\varphi_o(t)^2\,(|\,l>_{\mu\ \mu}<x\,| + |\,x>_{\mu\ \mu}<l\,|) =$$

$$= \begin{pmatrix} 1-(1-\alpha_l)\,\varphi_o(t)^2 + \varphi_x(t)^2 & \sqrt{\alpha_l\,\alpha_{j'}}\;\varphi_o(t)^2 \\ \sqrt{\alpha_l\,\alpha_j}\;\varphi_o(t)^2 & \sqrt{\alpha_j\,\alpha_{j'}}\left(\varphi_o(t)^2 - \dfrac{\varphi_x(t)^2}{1-\alpha_l}\right) \end{pmatrix}. \tag{6.138}$$

The deviation from a pure state is described by the quantity

$$\sigma(t) = \frac{1}{2}\,[1 - \mathrm{Tr}(\rho_\mu(t)^2)] = (1-\alpha_l)\,\varphi_o(t)^2\,[1 - \varphi_o(t)^2 + \varphi_x(t)^2]$$

$$-\,\varphi_x(t)^2\,[1 - (1-\alpha_l)\,\varphi_o(t)^2 + \varphi_x(t)^2]\,, \tag{6.139}$$

which satisfies

$$\begin{cases} 0 \leq \sigma(t) \leq \dfrac{1}{4} \\ \sigma(0) = \sigma(\infty) = 0\,. \end{cases} \tag{6.140}$$

For $n = 2$, $\sigma(t)$ determines the entropy $S(t)$ uniquely,

$$S(t) = -\,\mathrm{Tr}[\rho_\mu(t)\ln\rho_\mu(t)] =$$

$$= \left(\frac{1}{2} - \sqrt{\frac{1}{4} - \sigma(t)}\right)\ln\frac{1}{\sigma(t)} - 2\sqrt{\frac{1}{4} - \sigma(t)}\,\ln\left(\frac{1}{2} + \sqrt{\frac{1}{4} - \sigma(t)}\right). \tag{6.141}$$

In this case the entropy satisfies

$$\begin{cases} 0 \leq S(t) \leq \ln 2\,; \\ S(0) = S(\infty) = 0\,. \end{cases} \tag{6.142}$$

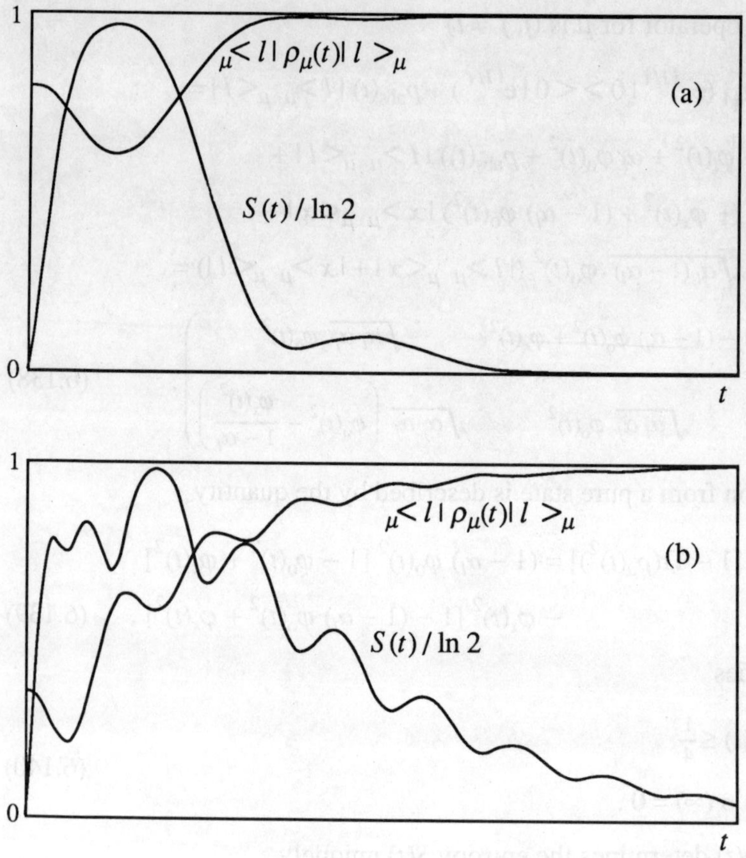

Fig. 6.4. Reduction of the wave packet according to the 'truncated model'. The probability $_\mu\langle l|\rho_\mu|l\rangle_\mu$ and the entropy $S(t)/\ln 2$ are plotted against time (arbitrary scale). (a) $\alpha_l = 0.80$, $\gamma = 2$; (b) $\alpha_l = 0.36$, $\gamma = 2$.

The time evolution of $_\mu\langle l|\rho_\mu(t)|l\rangle_\mu$ and $S(t)$ for $n = 2$ and some values α_l and γ is shown in Fig. 6.4.

Clearly this whole process is connected to a large entropy increase in M and its environment.

7. EVOLUTION AND VALUE

An evolutionary perspective is now an integral part not only of biology and geology, but also of physics through modern astrophysics and cosmology and through the theory of irreversible processes. In the concept of evolution there is the notion of *emergence*, that new features appear, which did not exist before. Implicit in this is also the notion of *value*. 'Higher' systems—systems of higher value—emerge out of lower systems, and value is in some sense being created.

'Value' is a concept not only in economics but also in philosophy. Value in the economic sense may be a means to increase value in the philosophical sense. For discussions on the value concept, see Hicks [1939] and Robinson [1978]. We saw in Section 2.4 that exergy as a general 'fuel' has a physical value, a concept very close to that of economic value. In this concluding chapter we shall start with a discussion of value in the simple physical/economic sense and then move towards the philosophical domain, all the time using the information theory approach.

7.1 Value of Exergy and Structure

We shall here tie physical value a little closer to the concept of instrumental value, as used in economics. First we shall consider a technical production system as the one discussed in Section 2.4. Then we shall apply economic aspects to self-organizing systems. If exergy in the form of chemical contrast is considered as a fuel or a raw material, a means to achieve structure, measured as structural contrast and regarded as valuable, then self-organization is a production process.

7.1.1 *Economic aspects of exergy*

At the end of Section 2.4 we stated that exergy can be given an "economic" interpretation as a measure of physical value. We shall here pursue this line of reasoning a little further. We thus consider an economy which is looked upon during a time interval short enough so that the economy may be considered as stationary, or, better, we consider averages over a certain time period. The constraints (2.174) may be given a rather general interpretation as:

availability constraints (for flow resources),
conservation constraints or
capacity constraints (for deposited resources),
environmental constraints (for emissions),
safety constraints.

In the stationary model represented by Fig. 2.6., the converter can now be interpreted as a combined production and distribution system [Eriksson 1984]. The extracted power $\dot{W}e$ is used to provide a flow of reasonably composed and distributed) *services* [Ayres 1978]. We assume the service mix to be of certain fixed proportions. The total service flow Γ is assumed to be variable but limited by the power $\dot{W}$,

$$\Gamma \leq \frac{1}{\lambda} \dot{W} \qquad (\lambda = \text{const.} > 0). \tag{7.1}$$

As is customary in economic theory, we introduce a *utility flow* $U(\Gamma)$. (For a discussion of this concept, see Samuelson [1983]; for the mathematical treatment, we also refer to Takayama [1974].) Its first derivative, the 'marginal utility', is assumed to be positive (more service is better) but decreasing,

$$\begin{cases} U'(\Gamma) > 0; \\ U''(\Gamma) < 0. \end{cases} \tag{7.2}$$

Our problem is stationary. To maximize the utility flow U we can now use an economic Lagrangian which contains the physical Lagrangian (2.183), here denoted L_{phys},

$$L_{\text{econ}} = U(\Gamma) - \gamma\lambda\Gamma + \gamma L_{\text{phys}}, \tag{7.3}$$

where γ is a Lagrange parameter, and where

$$\begin{cases} \gamma \geq 0; \\ \gamma(-\lambda\Gamma + L_{\text{phys}}) = 0. \end{cases} \tag{7.4}$$

The service production is said to be technically efficient, when (7.1) holds with equality. Since $L_{\text{phys}} = \dot{W}$, the second equation in (7.4) is satisfied independently of γ. We shall assume this to be the case and also assume that $\tau^* \neq 0$, i.e., that the second law constraint is effective. Then all results of Section 2.4 obtain, and

$$\frac{\mathrm{d}U(\Gamma)}{\mathrm{d}\Gamma} = \gamma\lambda\,. \tag{7.5}$$

From this and (7.1) with equality and from Section 2.4, Eq. (2.199) we then obtain the marginal utility of the various resource flows,

$$\frac{\partial U}{\partial P_m} = \frac{\mathrm{d}U}{\mathrm{d}\Gamma} \cdot \frac{1}{\lambda} \frac{\partial \dot{W}}{\partial P_m} = \gamma \left(\kappa^*_m - \tau^* \frac{\partial X(\mathbf{P}^*)}{\partial P^*_m} \right) . \tag{7.6}$$

Here γ can be identified as the marginal value of exergy, i.e., the utility increase per added unit of exergy. The generalized Carnot factor κ^*_m is the exergy content of the mth flow, and the last term in the bracket is the marginal exergy loss (through entropy production) per unit flow.

If the service flow is also limited by the flows of energy and materials, then the above discussion has to be slightly modified. Let us assume that

$$\boldsymbol{a}\Gamma \leq \sum_{j=1}^{M} P_j \boldsymbol{l}_j\,, \tag{7.7}$$

where $\boldsymbol{a}$ expresses the requirement of energy and materials, and where $\boldsymbol{l}_j$ expresses their technical availability in various flows. Then (7.7) may be taken into account through an extra term in the Lagrangian with a corresponding vector of Lagrange multipliers. The result of this after slight redefinitions is that the marginal utility of the *m*th flow, besides the terms (7.6) based on exergy, also obtains a term

$$\boldsymbol{p} \cdot \boldsymbol{l}_j \,, \tag{7.8}$$

where $\boldsymbol{p}$ is interpreted as an energy/materials *price* vector, reflecting scarcity in the restriction (7.7), i.e., those components of $\boldsymbol{p}$ vanish which correspond to strict inequality (no scarcity) in (7.7).

This discussion indicates how physical and economic concepts can be joined into an integrated theory. We now leave the economic terminology to discuss the use of energy and materials more in terms of physics and mathematical modelling.

A system which is limited in size and energy, has a finite number of states available. The time-development of such a system, whether it is due to internal dynamics, influence of noise or external control, or a combination of the three, can be thought of as a sequence of prints on the tape of a Turing machine. Thus within the physics of computation we can build models for the evolution of physical systems and for the action on physical systems.

Such models can be very appropriate for describing structure formation, where chemical contrast is transformed into structural contrast at the cost of entropy production. Clearly exergy as a single general resource measure is insufficient here. In the next part of this section, we shall describe a situation where structural contrast is valued higher than chemical contrast.

In particular, we could use a computer model for analysing the conversions of energy and materials in a society. The crucial function in this context is the minimum entropy production $X(\mathbf{P})$ of Section (2.4). As discussed in Section 4.5, a higher speed in the conversion processes may lead to higher entropy production. Precisely which dependence one has, is inherent in the technical methods, i.e., in the available algorithms for conversion of energy and materials.

It is clear that many algorithms for handling conversion processes lead to exergy losses due to friction, leakage, and dispersion of concentrated materials. A major demand for exergy supply in a society comes from the requirement that the lost exergy be replaced. A car which loses kinetic energy due to friction, gets it replaced from the exergy contained in the fuel. A house that leaks heat to its environment in winter, gets this heat replaced from electricity or fuel via the heating system. An exergy-rich material which is used for items that are worn out or dispersed through corrosion is replaced by the exergy-demanding processing industry.

In many cases techniques are known, by which friction, leakage, or dispersion could be reduced. But the incentives for such technological changes are weak, since many persons, companies, and nations have incomes proportional to the flows they are handling. Constraints of the kind mentioned in the beginning of this section are in the interest of all, but the incentives to work for the common good and to implement them may unfortunately be quite weak.

Moreover, erroneous or unclear descriptions of resource use, resource conversions, and technological options, often block the way for a broader understanding of the available possibilities. Economic processes in a society always involve matter, energy, entropy, and information. The aim of many economic activities is to achieve a certain structure, describable in information-theoretical terms. Still, it is commonly agreed that economists can deal with all this at any level, company, regional, national, or international, without any basic knowledge of thermodynamics or information theory. (For a physical approach to the description of economic systems and processes, see Ayres [1978], especially Chapter 3.) The flows of any economic activity must in some way be embedded in the natural flows of energy and materials on the surface of the earth. (For a discussion of economic growth under physical constraints, see Eriksson et al. [1984] and Månsson [1986].) Largely, economics is separated from the geophysical and biological sciences. Important efforts have been made, however, to bridge this gap [Clark 1976, Dasgupta and Heal 1979].

In the formal discussion in the beginning of this section, exergy played a major role. However, exergy is only the "fuel" for structuring matter or for maintaining structure. Even if the exergy of a structure is very small, it can still be quite valuable. Self-organizing systems are interesting and in this sense valuable not because of their exergy content, but because of their content of structural contrast.

7.1.2 *Build-up and maintenance of structure*

We shall here assume that structural contrast is more valuable than chemical contrast. Moreover we assume that apart from its stock value, structural contrast represents a flow value, i.e., a value per unit time that it exists. Let us consider again the closed system of Fig. 5.6a (Section 5.2).

If the stock values per unit of structural and chemical contrast are p_{str} and p_{chem}, then the total stock value Ω at time t is

$$\Omega(t) = p_{\text{chem}} K_{\text{chem}}(t) + p_{\text{str}} K_{\text{str}}(t) \,. \tag{7.9}$$

If the value flow per unit of structural contrast is π_{str}, then the total amount of value produced per unit time is (Fig. 7.1)

$$\omega(t) = \dot{\Omega}(t) + \pi_{\text{str}} K_{\text{str}}(t) \,. \tag{7.10}$$

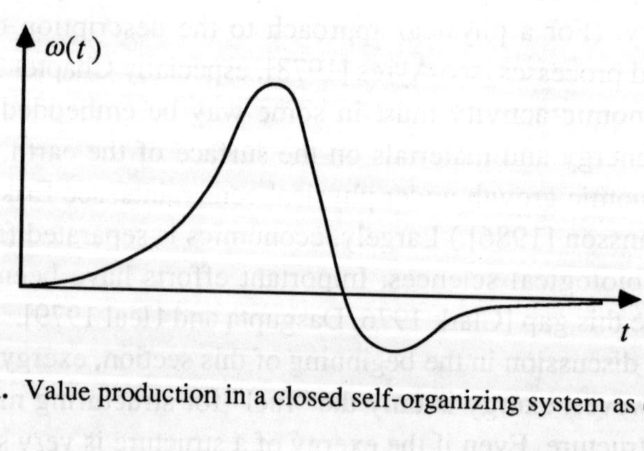

Fig. 7.1. Value production in a closed self-organizing system as described by (7.10).

The total (net) value produced in the whole process is

$$\Omega_{\text{tot}} = \int_0^\infty \mathrm{d}t\ \omega(t) = \pi_{\text{str}} \int_0^\infty \mathrm{d}t\ K_{\text{str}}(t) - p_{\text{chem}} K_{\text{chem}}(0)\ . \tag{7.11}$$

For an open system (cf. Fig. 5.6b), we have to subtract from (7.10) the cost in terms of the net flow $\Phi_{\text{chem}}(t)$ into the system of chemical exergy, that has to be provided from the environment. Thus the value produced per unit time is

$$\omega(t) = \dot{\Omega}(t) + \pi_{\text{str}} K_{str}(t) - p_{\text{chem}} \Phi_{\text{chem}}(t)\ . \tag{7.12}$$

If a stationary state obtains, then

$$\omega = \pi_{\text{str}} K_{str} - p_{\text{chem}} \Phi_{\text{chem}}\ , \tag{7.13}$$

which may be compared to the original (unstable) unstructured stationary state indicated by a star,

$$\omega^* = - p_{\text{chem}} \Phi^*_{\text{chem}}\ . \tag{7.14}$$

In the difference between (7.13) and (7.14),

$$\omega - \omega^* = \pi_{\text{str}} K_{str} + p_{\text{chem}} (\Phi^*_{\text{chem}} - \Phi_{\text{chem}})\ , \tag{7.15}$$

it is not unusual that both terms, the structure term and the relative cost term, are positive.

The discussion here shows that it is futile to try to reduce all aspects of value to one standard, such as exergy. Exergy is necessary to build structure, but structural constrast is more than exergy.

7.1.3 *Textual contrast and correlational contrast*

In Section 3.3 we decomposed structural contrast into correlational and textual contrast. We shall now leave the more formal economics-inspired considerations and discuss the characteristics of correlational contrast and textual contrast, and the question in what way they are valuable in their actual manifestations.

Whereas the correlational contrast of a system contains the information—present everywhere in the system—on the typical features of the system, the textual contrast describes the local variations against this background of typical features. The information contained in the correlational contrast can be used to identify the system, as compared to other systems, by identifying its inner structure. Because of the translational invariance of the correlations, it can not be used for determining the localization within the system. For this one uses the information contained in the textual contrast, provided it stays constant or changes regularly during a sufficiently long time interval. The textual structure at the microscopic level does not possess any such permanence. In fact, at this level we have to use a statistical mechanics description. In an ensemble of systems at the scale of the fluctuations, there is no way to read off the information in the textual structure. The ensemble average of the microscopic textual contrast is the entropy (see Section 4.4), and the flow of contrast into the microscopic domain is the entropy production (Section 3.2).

The information contained in correlational contrast and in textual contrast to the extent that they possess the necessary permanence thus determine *context*. The correlations help to *discriminate one context from another*; the information contained in the textual contrast helps to *localize a position within a context*. (When coming to an unknown city, one can manage quite well by using one's knowledge of how similar cities are usually structured. But to find a given address one has to look for particular objects that may be identified against the familiar general structure.)

The structure of a material (granular, fiber, etc [Ayres 1978]) is contained in its correlational contrast; the shape of an artificially made object, such as a car or a telephone, lies in its textual contrast. In both cases there is a requirement to conform within certain tolerance limits to a prescribed ideal.

Textual contrast in various forms is used to carry messages. Besides the requirement of permanence, one also requires fidelity to the original or the intended message. Redundancy, i.e., correlational contrast is a means to diminish the effect of errors. Redundancy is on the whole confidence-building, reassuring, but it may also be boring.

In the arts, the tension between textual contrast and correlations, between the individual and the typical is a major theme on various levels: within the same painting, novel or piece of music itself, or in the relation between one piece of art and the tradition which it belongs to or opposes.

In order not to get lost in arbitrary structures at the microscopic level, one can use an ensemble to describe the system, as discussed above, or one can stop the analysis at a certain scale length (resolution) and consider all finer details as irrelevant. In this way correlational and textual contrast become useful also as measures of complexity.

7.1.4 *Algorithmic information*

It has been the ambition in the previous chapters to use, as far as possible, concepts that are rather easily determined from the state of the system. Correlational contrast and textual contrast are concepts of this kind. The algorithmic information is not so easily determined. Sometimes it can be approximated by textual contrast.

Still, when the computer time is not prohibitively long (i.e., the depth is not prohibitively large), the algorithmic information is in some sense a true reduced measure of complexity. Algorithimic information may have quite a small significance at a purely physical level; at the level of biochemistry and molecular biology, nature has built functioning Turing machines, and algorithmic information is a very relevant concept. As we mentioned in Section 5.3, the structural information may differ very little between living and non-living systems. Therefore, to characterize life, we really need more sophisticated concepts than correlational and textual contrasts.

The restrictions made in Chapters 3 and 4 to systems for which the thermodynamic limit exists, cannot in general be applied to biological systems. Here one needs concepts that do not require large-scale homogeneity. The copying of DNA and the synthesis of proteins are fundamental processes in any living organism, which can be described as a Brownian computer. It then seems reasonable to use algorithmic information as a basic concept. It may even be so that the difficulty to determine when an algorithm is optimal reflects an important feature of living systems.

As a measure of structural organization within a system, Chaitin [1979] suggested the mutual algorithmic information of its subsystems. By varying the size of the subsystems, one can determine at what scale the structural organization is located. The algorithmic redundancy used in Section 4.2.4 is a slight modification of Chaitin's concept, defined to be easily applicable, first to cellular automata and further to more general systems.

Cytochrome-C is a protein of the cell nucleus, which has been used to study the relative closeness and hence systematics of living species at the molecular level. Ebeling and Jiménez-Montaño [1980] have studied information-theoretical aspects of cytochrome-C. One-dimensional lattices such as cellular automata can be described in linguistic terms such as Chomsky's [1956] grammar concepts, which were used by Ebeling and Jiménez-Montaño. The evolutionary tree as reflected in the hierarchy of cytochrome-C can then be described as a tree of (so-called context sensitive) grammars that have become increasingly restricted with progressing evolution.

In the study of lattices, also multi-dimensional ones, a hierarchical classification appears as natural [Parisi 1987]. Hierarchies of structures (e.g., [Huberman and Hogg 1986]) are certainly fields of great relevance for the theme of this book, but it has not been possible for us to include them in our discussion.

7.2 An Evolutionary Perspective

We now return to the evolutionary perspective, in the Introduction presented as a theme of the book. In this and the following section we shall review some concepts and ideas of the book in relation to this theme. The theme itself is great and inspiring. The concepts and ideas are sufficient only for the lowest—but all the more fundamental—steps of the evolutionary ladder.

7.2.1 *Cosmic, biological, and societal evolution*

The emerging picture of the universe is that it came into being, together with space-time itself, out of nothing. Like there is no question of "outside" the universe in space, there is also no question of what was before the start of the universe, the Big Bang.

The transition from chemical exergy into structural exergy was discussed earlier in this chapter as a process in which value is created—if structural exergy can be viewed as more valuable than mere chemical exergy.

The idea that something valuable arises out of something with smaller value is inherent in the idea of evolution itself, that "higher forms" evolve out of "lower." The first step in the evolution process was thus the step from nowhere and never into space-time and from nothingness into existence. It may be quite natural to think of *existence* as the most primitive of values.

One feature of the emerging picture is that radiation and high temperature matter (together with antimatter) was "bought" by the universe, during a phase of rapid expansion, inflation, at the expense of a negative energy per unit volume of the growing space. If the matter/antimatter energy is considered more valuable than the field energy of volume itself, then the inflationary phase also involves creation of value.

There may be other phases in the very early universe which for similar reasons can be viewed as value-creating. We focused in Section 2.3 on a process which took place later, after the antimatter had annihilated and the excess matter from an earlier imbalance was present in the form of nucleons. The temperature of the nucleon gas was determined by the radiation, which cooled during the expansion. Later the nucleon gas left thermal equilibrium and became supercooled, since its equilibrium state became a state of heavier nuclei, but the possibilities for nuclear reactions were not favourable, and only helium was formed. In this way, during the era of seconds to hours after the Big Bang, the nuclear exergy was formed, which much later came into use as the fuel of the stars. In particular this is the fuel of the Sun, and hence of life on Earth and of human activities. Some of the fuel used by human societies is not taken from the Sun directly, but has been accumulated on the Earth in the chemical non-equilibrium between the oxygen of the atmosphere and the hydrogen and hydrocarbons of the fossile fuels beneath the surface of the Earth.

The value of the cosmic nuclear exergy lies not only in its existence, but also in the fact that it has been made available by nuclear reactors, i.e., the stars. They transform it into radiation of a few thousand degrees, available to water molecules and to the rich variety of carbon compounds. Stars and planets owe their existence to the inherent instability of gravitational attraction. When the universe had cooled below ionization temperature for hydrogen and helium, neutral atoms formed and the coupling between matter and radiation almost disappeared. The pressure felt by matter dropped drastically, and the gravitational forces were free to clump matter together. This represents another step of exergy creation, the creation of gravitational exergy. Actually, gravitational exergy may be considered as being continuously created from that time on, as matter is "lifted" apart through the expansion and thus gaining potential energy.

Matter was clumping together on several scales, not only into stars and planets, but also into clusters of stars, galaxies, and clusters of galaxies. The clumping of matter was counteracted by the expansion; the distances that had to be overcome were themselves expanding. Dyson [1971] has emphasized the delays, "hang-ups", in these processes, first due directly to the distances involved, secondly due to angular momentum and centrifugal forces, statistically inevitable when a large cloud contracts. In the stars, a very important hang-up is that the nuclear reactions are slow since they involve weak interaction.

The stellar hang-ups keep the stars in a very long-lived state of near stationary radiation. Such permanence is a necessary condition for the evolution of life. Permanence, within certain margins in several geophysical variables, is also a prerequisite for life. On the other hand, changing conditions may be necessary for changing the selection criteria to give room for new evolutionary pathways. To the dichotomy between permanence and change there is a corresponding dichotomy within life itself, that of universality and variation. Thus, we meet again the tension between the typical and the unique, between correlational and textual information.

Our discussion on life and evolution has been extremely limited. We have only discussed in Section 4.6 the evolutionary measures introduced in ecology. In Section 5.3 we have mentioned the hypercycles as a crucial evolutionary step, and we have shown that the evolutionary measures can be applied to competing hypercycles.

7.2.2 *Self-reference and selves*

Each cell in a multicellular organism carries, in the DNA of the chromosomes, algorithms for producing not only the cell itself but the whole variety of cells making up the organism. The cell or the organism carries algorithmic information about itself. In terms of Russell's theory of logical types, the metalevel describing the system is part of the system itself. This *self-reference* is thus a basic characteristic of life itself [Bateson 1972, Hofstadter 1980].

This self-reference is not complete, in the sense that it requires a certain stationary or periodic physico-chemical environment. One can say that it involves the information modulo such an environment. If one would like to express this in terms of the Turing machine, one could say that the physico-chemical properties of the environment are internal features of the Turing machine itself.

This possibility of an abstract representation of self-reference shows that what matters are the systemic properties and the fact that they are materially manifested in some way, not the specific way chosen for this manifestation. An information-theoretical theory of self-reference still seems to be lacking.

Self-reference in the individual's collection of information and in the choice between different ways of action is probably an essential feature of consciousness. The systemic view is strengthened by the experiences that a split-brain patient who gets different information inputs in the right and left halves of his brain, also shows a split consciousness. On the other hand, persons working intensively on a common task with very frequent communication can act as if having a joint consciousness.

A step beyond or above that of consciousness is self-consciousness: a conscious being reflecting over itself as a conscious being. ('It' then turns into a person and 'itself' becomes 'herself' or 'himself'.)

Self-consciousness may also exist on various levels. A small strongly integrated group of the kind that we mentioned may develop a strong feeling of group identity. There is not only a common consciousness but also a common self and a common self-consciousness.

Larger selves exist in most human societies, characterized by common ways of thinking, common feelings, and a sense of identity involving mutual respect and mutual responsibility between the members.

A major problem for humankind of today is the following. On one hand, we have a global civilization: the technology and administrative practices of the industrialized societies are spreading all over our world. This involves radical changes, and practically all societies and all ecosystems are affected, even the biosphere itself. On the other hand, the powerful social actors belong to limited selves: nations, companies, interest groups. Most people have no influence on the global development, and the development is more for the benefits of the powerful actors than for the people of the world. In this situation we lack a global human identity with due respect and responsibility for individuals, for cultures and societies and for the heritage that we have in common with our fellow living beings, the biosphere. We are still lacking a strong common self to defend human values and the values of life.

Common ways of thinking, common feelings and a human identity would mean a new global culture. To be true and to be viable, such a culture must have scientific knowledge and scientific inquiry as essential cognitive elements.

7.3 Concluding remarks

Music cannot be transmitted to us without matter. Thoughts cannot be thought without a brain. Spiritual life needs matter.

The mind may influence and even control material events, but the mind itself needs matter and energy for its very existence. It needs more; it needs non-equilibrium matter/energy. We can go on: the material system carrying the mind needs more: it needs structure, internally correlated as well as with reference to the outside world (context). The actions of this system must be coordinated (organization). All this requires a certain complexity/depth. To carry a mind, the system must probably also refer to itself. Self-reference may be needed on several levels. This book has dealt only with a few of the very simplest levels, the lower part of the ladder in Fig. 7.2.

- Self-knowledge
- Knowledge
- Self-consciousness (on various levels)
- Consciousness
- Know-how
- Life
- Complexity
- Organization
- Structure/Context
- Contrast
- Matter/Energy in space-time

Fig. 7.2. Ladder of evolution/value.

Our conclusion is that on the levels that we have studied, information theory works and provides useful concepts. We believe that it will do so also on the higher levels, which are more difficult to analyse. Maybe this has happened already—important work may exist, but we may have failed to see the proper connection to our perspective and our questions.

A useful theoretical tool in this context is the Turing machine. In the real world practically all systems are subject to some kind of stochastic interaction. Hence, we should use a *non-deterministic* Turing machine. Moreover, the system which this Turing machine is going to model is also a physical system, subject to the laws of thermodynamics. The physical restraints on computation, discussed in Section 4.5, should therefore be incorporated into the model. If we let the term 'thermodynamic' refer not only to this, but also to the presence of stochastic influence, then the proper tool to use is a *thermodynamic Turing machine*.

We have expressed a belief that information theory has much more to give than what we have been able to discuss in this book. If we look into the subjects dealt with in the social sciences and the humanities, we find that they deal to a very large extent with human communication: how persons, groups or cultures express themselves, how messages are conveyed in the society, how they are interpreted or could be interpreted, interaction between persons or groups through exchange of information. If new concepts are formed within information theory, they may be relevant also to these questions. Therefore, information theory carries a potential to bridge the gap between the 'two cultures' [Snow 1964], on one hand the natural sciences, mathematics, and technology, on the other the social sciences and the humanities. The old question of meaning belongs here. Life is meaningful, because it offers the possibility to create meaning.

REFERENCES

Ahern, J.E., 1980, *The Exergy Method of Energy Systems Analysis*, John Wiley & Sons, New York.

Andresen, B., Salamon, P., and Berry, R.S., 1977, "Thermodynamics in finite time: Extremals for imperfect heat engines," J. Chem. Phys. **66**, 1571.

Ash, R., 1965, *Information Theory*, Interscience, New York.

Ayres, R.U., 1978, *Resources, Environment, and Economics*, John Wiley & Sons, New York.

Bateson, G., 1972, *Steps to an Ecology of Mind*, Ballantine Books, New York.

Benioff, P., 1982, "Quantum mechanical models of Turing machines that dissipate no energy," Phys. Rev. Lett. **48**, 1581.

Bennett, C.H., 1973, "Logical reversibility of computation," IBM Journal of Research and Development **17**, 525.

Bennett, C.H., 1982, "The thermodynamics of computation—a review", Int. J. Theor. Phys. **21**, 905.

Bennett, C.H., 1986, talk given at the conference "Physics of Complexity" at ICTP, Trieste, Italy, 1-5 September 1986.

Berry, R.S., 1972, Bulletin of the Atomic Scientists **9**, 8.

Blumenfeld, L.A., 1981, *Problems of Biological Physics*, Springer-Verlag, Berlin.

Boulding, K.E., 1978, *Ecodynamics*, Sage Publications, Beverly Hills.

Brillouin, L., 1956, *Science and Information Theory*, Academic Press, New York; see also second edition 1962.

Brønsted, J.N., 1955, *Principles and Problems of Energetics*, Interscience, New York.

Brudno, A.A., 1977, "On the complexity of trajectories in dynamical systems," Usp. Mat. Nau. **33**, 207. (In Russian.)

Carnot, N.L.S., 1824, *Réflections sur la puissance motrice du feu et sur les machines propres a développer cette puissance*, Bachelier, Paris; R. Fox, ed., Libraire Philosophique J. Vrin, Paris 1978.

Chaitin, G.J., 1966, "On the length of programs for computing finite binary sequences," Journal of the Association for Computing Machinery **13**, 547.

Chaitin, G.J., 1975, "A theory of program size formally identical to information theory," Journal of the Association for Computing Machinery **22**, 329.

Chaitin, G.J., 1979, "Toward a mathematical definition of 'life'," in [Levine and Tribus 1979], p. 477.

Chomsky, N., 1956, "Three models for the description of language," IRE Trans. on Information Theory **2**, 113.

Clark, C.W., 1976, *Mathematical Bioeconomics*, John Wiley & Sons, New York.

Cvitanovic, P., 1984 (ed.), *Universality in Chaos*, Adam Hilger Ltd, Bristol, 1984.

Dasgupta, P.S. and Heal, G.M., 1979, *Economic Theory and Exhaustible Resources*, Cambridge University Press, Cambridge.

Davies, P.C.W., 1974, *The Physics of Time Asymmetry*, University of California Press, Berkeley.

Dyson, F.J., 1971, "Energy in the universe," Scient. Am. **225**(3), 50.

Ebeling, W. and Jiménes-Montaño, M.A., 1980, "On grammars, complexity, and information measures of biological macromolecules," Mathematical Biosciences **52**, 53.

Eckman, J.-P. and Ruelle, D., 1985, "Ergodic theory of chaos and strange attractors," Rev. Mod. Phys. **57**, 617.

Eigen, M. and Schuster, P., 1979, *The Hypercycle*, Springer-Verlag, Berlin.

Epstein, I.R., 1984, "Complex dynamical behavior in 'simple' chemical systems," J. Phys. Chem., **88**, 187-198.
Eriksson, K.-E., 1984, "Thermodynamical aspects on ecology/economics," in Jansson, A.-M. (ed.), *Integration of Economy and Ecology: An Outlook for the Eighties* , Wallenberg Symposium, Stockholm, p. 39.
Eriksson, K.-E., 1985a, "Decomposition of information with respect to position and structure size," Physica Scripta **T9**, 133.
Eriksson, K.-E., 1985b, "Random walk in a simplex—the way 'God plays dice'?" Physica Scripta **32**, 277.
Eriksson, K.-E., 1986, "Decomposition of correlation information with respect to correlation length and scale," Physical Resource Theory Group report no. 86-8.
Eriksson, K.-E., 1987, "Quantum statistics of quantum measurement," Physica Scripta **35**, in press.
Eriksson, K.-E., Islam, S., and Skagerstam, B.-S., 1982, "A model for the cosmic creation of nuclear exergy," Nature **296**, 540. An error in this paper has been corrected in the present text, leading to slightly modified results. See this paper for further references.
Eriksson, K.-E. and Kåberger, T., 1986, "Flow-based measures of system evolution," Physical Resource Theory Group report no. 86-2.
Eriksson, K.-E. and Lindgren, K., 1987, "Structural information in self-organizing systems," Physica Scripta **35**, 388.
Feynman, R.P., 1985, "Quantum mechanical computers," Optics News **11**, 11.
Field, R.J., and Burger, M. (eds.), 1984, *Oscillations and Travelling Waves in Chemical Systems*, John Wiley & Sons, New York.
Frautschi, S., 1982, "Entropy in an expanding universe," Science **217**, 593.
Fredkin, E. and Toffoli, T.,"Conservative logic," 1982, Int. J. Theor. Phys. **21**, 219.
Fujisaka, H., 1983, "Statistical dynamics generated by fluctuations of local Lyapunov exponents," Prog. Theor. Phys. **70**, 1264.
Gaggioli, R.A. (ed.), 1980, *Thermodynamics: Second Law Analysis* , ACS Symposium Series 122, American Chemical Society, Washington.
Gaggioli, R.A. (ed.), 1983, *Efficiency and Costing* , ACS Symposium Series 235, American Chemical Society, Washington.
Gamow, G., 1965, *Mr Tompkins in Paperback*, Cambridge University Press, p. 91.
Gardner, M., 1971, "Mathematical games," Sci. Amer. **224**(2), 112; (3), 106; (4), 114.
Georgescu-Roegen, N., 1971, *The Entropy Law and the Economic Process*, Harvard University Press, Massachusetts.
Gibbs, J.W., 1873, *Collected Works*, Vol. I, p. 53, Yale University Press, New Haven 1948; Originally published in Trans. Conn. Acad., Vol. II, pp. 382-404.
Gierer, A. and Meinhardt, H., 1972, Kybernetik **12**, 30. (We have used a slightly modified model.)
Gisin, N., 1984, "Quantum measurement and stochastic processes," Phys. Rev. Lett. **52**, 1657.
Glansdorff, P. and Prigogine, I., 1971, *Thermodynamic Theory of Structure, Stability and Fluctuations*, John Wiley & Sons, London.
Glauber, R.J., 1963, "Coherent states and incoherent states of the radiation field," Phys. Rev. **131**, 2766.
Gödel, K., 1931, "Über formal unentschiedbare Sätze der Principia Mathematica und verwandter Systeme, I", Monatshefte für Math. und Physik. English translation in Davies, M., *The undecidable*, Raven Press, 1965.
Grassberger, P., 1986a, "Towards a quantitative theory of self-generated complexity," Int. J. Theor. Phys. **25**, 907.

Grassberger, P., 1986b, "Long-range effects in an elementary cellular automaton," J. Stat. Phys. **45**, 27.
Grassberger, P. and Procaccia, I., 1982, "Characterization of strange attractors," Phys. Rev. Lett. **50**, 346.
Grassberger, P. and Procaccia, I., 1984, "Dimensions and entropies of strange attractors from fluctuating dynamics," Physica **13D**, 34.
Grimmett, G. and Stirzaker, D., 1982, *Probability and Random Processes*, Clarendon Press, Oxford.
Grubbström, R.W., 1980, *Towards a Theoretical Basis for Energy Economics*, Naval Postgraduate School, Monterey, Calif., Techn. Rep. NPS-54-80-015.
Gurel, O. and Gurel, D., 1983, *Oscillations in Chemical Reactions*, Springer-Verlag, Berlin.
Haken, H., 1977, *Synergetics*, Springer-Verlag, Berlin.
Haken, H. (ed.), 1980, *Synergetic systems*, Springer-Verlag, Berlin.
Haken, H., 1983, *Synergetics*, 3rd ed., Springer-Verlag, Berlin.
Haken, H., 1984, *Advanced Synergetics*, Springer-Verlag, Berlin.
Hayes, B., 1983, "A progress report on the fine art of turning literture into drivel," Scient. Am. **249**(5), 16.
Hentschel, H.G.E. and Procaccia, I., 1983, "The infinite number of generalized dimensions of fractals and strange attractors," Physica **8D**, 435.
Herzel, H., Ebeling, W., and Schulmeister, Th., 1987, "Nonuniform chaotic dynamics and effects of noise in biochemical systems," Z. Naturforsch. **42a**, 136.
Hicks, J.R., 1939, *Value and Capital*, Oxford University Press.
Hiernaux, J. and Babloyantz, A., 1976, "Dissipation in embryogenesis," J. Non-Equilib. Thermodyn. **1**, 33.
Hofstadter, D.R., 1980, *Gödel, Escher, Bach: an Eternal Golden Braid*, Vintage Books, New York.
Hongler, M.-O., 1979, "Exact solutions of a class of non-linear Fokker-Planck equations," Phys. Lett. **75A**, 3.
Hopcroft, J.E., 1984, "Turing machines," Scient. Am. **250**(5), 70.
Hopcroft, J.E. and Ullman, J.D., 1979, *Introduction to Automata Theory, Languages, and Computation*, Addison-Wesley Publishing Company, Reading, Massachusets.
Hubernman, B.A. and Hogg, T., 1986, "Complexity and adaptation," Physica **22D**, 376.
Jäger, W. and Murray, J.D. (eds.), 1984, *Modelling of Patterns in Space and Time*, Lecture Notes in Biomathematics 55, Springer-Verlag, Berlin.
Jaynes, E.T., 1957, "Information theory and statistical mechanics," Phys. Rev. **106**, 620.
Jaynes, E.T., 1963, "Information theory and statistical mechanics," in Ford, K.W. (ed.), *Statistical Physics*, Brandeis Lectures, 1962.
Jaynes, E.T., 1979, "Where do we stand on maximum entropy?", in [Levine and Tribus 1979], p. 15.
Jaynes, E.T., 1983, *Papers on Probability, Statistics and Statistical Physics*, R.D. Rosenkrantz, ed., D. Reidel, Dordrecht, Holland.
Johansson, A.E.I., Peresutti, G., and Skagerstam, B.-S., 1982, "The big bang nucleosynthesis and finite temperature field theory," Phys. Lett. **117B**, 171.
Kaneko, K. and Akutsu, Y., 1986, "Phase transitions in two-dimensional stochastic cellular automata," Journal of Physics A **19**, L69.
Karlsson, S., 1982, "The exergy of incoherent electromagnetic radiation," Physica Scripta **26**, 329.
Kinzel W., 1985, "Phase transitions of cellular automata," Z. Physik B **58**, 229.

Klauder, J.R. and Skagerstam, B.-S. (eds.), 1985, *Coherent States*, World Scientific, Singapore.
Kneese, A.V., Ayres, R.U., and d'Arge, R.C., 1970, *Economics and the Environment*, Resources for the Future, Inc., Washington, D.C.
Kolmogorov, A.N., 1965, "Three approaches to the quantitative definition of information", Prob. Peredach. Inform. **1**, 3. (In Russian.)
Kullback, S., 1959, *Information Theory and Statistics* , Wiley, New York.
Küppers, B.-O., 1983, *Molecular Theory of Evolution*, Springer-Verlag, Berlin.
Landauer, R., 1961, "Irreversibility and heat generation in the computing process," IBM J. Res. Dev. **5**, 183.
Landauer, R., 1987, "Computation—a fundamental physical view," Physica Scripta **35**, 88.
Landsberg, P., 1978, *Thermodynamics and Statistical Mechanics*, Oxford University Press.
Layzer, D., 1975, "The arrow of time," Scient. Am. **233**(6), 56.
Leontief, W., 1951, *The Structure of American Economy, 1919-1939*, Oxford University Press, New York (2nd ed.).
Leung-Yan-Cheong, S.K. and Cover, T.M., 1978, "Some equivalences between Shannon entropy and Kolmogorov complexity," IEEE Trans. Inform. Theory **24**, 331.
Levine, R.D. and Tribus, M. (eds.), 1979, *The Maximum Entropy Principle*, MIT Press, Cambridge, Massachusetts.
Lind, D.A., 1984, "Applications of ergodic theory and sofic systems to cellular automata," Physica **10D**, 36.
Lindblad, G., 1983, *Non-Equilibrium Entropy and Irreversibility*, Reidl Publishing Company.
Lindgren, K., 1987, "Correlations and random information in cellular automata," Complex Systems **1**, 529.
Lindgren, K. and Månsson, B.Å., 1986, "Entropy production in a chaotic chemical system," Z. Naturforsch. **41a**, 1111.
Mandelbrot, B.B., 1983, *The Fractal Geometry of Nature*, WH Freeman and Company, New York.
Månsson, B.Å., 1985, "Entropy production in oscillating chemical systems," Z. Naturforsch. **40a**, 877.
Månsson, B.Å., 1986, "Optimal development with flow-based production," Resources and Energy **VIII**, 109.
Martin-Löf, P., 1966, "The definition of random sequences", Inform. Contr. **9**, 602.
Maxwell, J.C., 1875, *The theory of heat*, 4th ed., Longmans, Green, and Co., London.
Messiah, A., 1961, *Quantum Mechanics*, North-Holland Publishing Company, Amsterdam.
Miller, J.G., 1978, *Living Systems*, McGraw-Hill, New York.
Moran, M.J., 1982, *Availability Analysis: A Guide to Efficient Energy Use*, Prentice-Hall, Englewood Cliffs, New Jersey.
Nicolis, G. and Prigogine, I., 1977, *Self-Organization in Non- Equilibrium Systems*, John Wiley & Sons, New York.
Nicolis, J.S., Mayer-Kress, G., and Haubs, G., 1983, "Non-uniform chaotic dynamics with implications to information processing," Z. Naturforsch. **38a**, 1157.
Odum, H.T., 1971, *Environment, Power, and Society*, John Wiley & Sons, New York.
Oparin, A.I., 1961, *Life: its nature, origin and development*, Academic Press, New York.